T0212856

The Plant-Parasitic Nematode Genus *Meloidogyne* Göldi, 1892 (Tylenchida) in Europe

BY

Gerrit Karssen

BRILL

LEIDEN · BOSTON · KÖLN

2002

This book is printed on acid-free paper.

Library of Congress Cataloging-in-Publication Data

Library of Congress Cataloging-in-Publication Data is also availalbe

Die Deutsche Bibliothek - CIP Einheitsaufnahme

The plant-parasitic nematode genus Meloidogyne Göldi, 1892 (Tylenchida) in Europe / Gerrit Karssen. - Leiden ; Boston ; Köln : Brill, 2002

ISBN 90 04 12790 9

Root knot-A nematode-caused disease characterized by round to irregular galls (knots) on the roots, caused by *Meloidogyne* species. Most common in sandy soils or untreated greenhouse beds, attacking over 2.000 kinds of plants.

Glossary of Plant-Pathological Terms (APS), 1997.

CONTENTS

1

INTRODUCTION

"Öb Gemische von *Meloidogyne*-Arten öfters vorkommen, ist
zur Zeit nicht zu beurteilen. Die Indentifikation der einzelnen
Arten ist erst seit kurzem bekannt und stützt sich auf variable
morphologische Merkmale, sodass man geneigt ist, abweich-
ende Stücke an erster Stelle als Varianten innerhalb der Art
zu betrachten."

M. Oostenbrink (1957).

Root-knot nematodes?

The genus *Meloidogyne* Göldi, 1892, or root-knot nematodes, represent a relatively small
but economically important group of obligate plant pathogens. They are distributed world-
wide and parasitize on almost every higher plant species. While reproducing and feeding
within roots, they induce galls or root-knots and disorder the physiology of the infected plant,
reducing crop yield and product quality (Eisenback and Hirschmann, 1991; Jepson, 1987).

More than eigthy nominal species have been described so far (Appendix I), about ten
species are agricultural pests, four are major pests and distributed world-wide in agricultural
areas, two species were recently added to the European list of quarantine organisms, to
prevent further distribution within Europe.

Classification

Classification of the plant-parasitic nematode genus *Meloidogyne* Göldi, 1892, within the
Animalia. Partly after the classification of the suprakingdom Animalia *sensu* Möhn (1984),
phylum Nematoda *sensu* Maggenti (1991) and the suborder Tylenchina *sensu* Maggenti *et al.*
(1987):

```
Suprakingdom  Animalia (Metazoa)
    Kingdom  Bilateralia
        Superphylum  Pseudocoelia
            Midphylum  Nemathelminthes
                Phylum  Nematoda
                    Class  Secernentea
                        Subclass  Diplogasteria
                            Order  Tylenchida
                                Suborder  Tylenchina
                                    Superfamily  Tylenchoidae
                                        Family  Heteroderidae
                                            Subfamily  Meloidogyninae
                                                Genus  Meloidogyne Göldi, 1892
```

The present position of the genus *Meloidogyne* within the family Heteroderidae Filipjev & Schuurmans Stekhoven, 1941 *sensu* Luc *et al.* (1988) is uncertain. Based on a wide range of morphological data, Baldwin (1992) reported, it is likely that Heteroderinae are not monophyletic with Meloidogyninae. Recently Geraert (1997) demonstrated that Meloidogyninae has the same head end-on view as the Pratylenchidae, while Heteroderinae has the same head as the Hoplolaimidae. We may expect some changes in the systematical composition of the family Heteroderidae in the nearby future.

Genus description

Genus *Meloidogyne* Göldi, 1892 (*Meloidogyne* means 'apple shaped female').
Type species: *M. exigua* (Göldi, 1892) Chitwood, 1949.

Female

Sedentary, white, rounded to pear-shaped body with protruding neck, cyst stage absent. Ranging in length from 350 μm to 3 mm. Cuticle annulated, posteriorly with a characteristic unique pattern around the perineum: the perineal pattern. Anus and vulva terminal, phasmids near the anus, usually anus covered with cuticular fold, perineum sometimes slightly elevated.

Head not or slightly set off, cephalic frame work distinct but delicate. Labial disc not to slightly raised, fused with medial and lateral lips. Two slit-like amphidial and ten small sensilla openings present. Delicate stylet, ranging in length from 10-25 μm, cone in most species slightly curved dorsally, shaft straight with three knobs. Dorsal pharyngeal gland orifice (DGO) ranging from 2.5-9.0 μm behind the knobs.

Secretory-excretory pore (S-E pore) located between head end and metacorpus level. Metacorpus relatively large, posteriorly with pharyngeal glands, variable in size and shape, ventrally overlapping the intestine.

Long didelphic, partly convoluted, gonads present. Most of the unembryonated eggs deposited in an egg-sac, formed by six large rectal glands and secreted through the anus.

Male

Vermiform, non-sedentary, annulated and ranging in length between 600-2500 μm. Head composed of head cap and head region (= post-labial annule). Head cap with rounded labial disc and four fused medial lips. Six inner labial sensilla centred around the stylet stoma and one cephalic sensillum present on each medial lip.

Large slit-like amphidial openings located between labial disc and lateral lip. Lateral lips in some species reduced or absent. Head region sometimes set off and/or partly subdivided by transverse incisures. Cephalic framework and straight stylet well developed, the latter ranging in length between 13-33 μm. DGO located 2-13 μm behind the three stylet knobs.

Metacorpus smaller than in females. S-E pore and hemizonid located between metacorpus level and the ventrally overlapping of the pharyngeal glands. Hemizonid anterior or sometimes posterior to S-E pore. Pharyngeal glands nuclei reduced to two.

One long testis usually present, rarely two reduced ones. Lateral field in most species with four incisures, outer bands often areolated.

Tail short, bluntly rounded, without a bursa, phasmids near cloaca. Spicules slender, 20-40 μm long; gubernaculum crescentic, about 10 μm long.

Second-stage juvenile

Vermiform, infective stage, annulated and ranging in length between 250-600 μm. Head structure as in males, but much smaller and with weakly sclerotized cephalic framework. Delicate straight stylet, about 9-16 μm long. DGO 2-12 μm behind the knobs.

Metacorpus relatively small. Hemizonid anterior or posterior to the S-E pore. Three pharyngeal glands present, ventrally overlapping the intestine.

Rectum often inflated. Tail 15-100 μm long, tapering towards hyaline tail part. Lateral field with four incisures.

Third and fourth-stage juvenile sedentary inside root and swollen, no stylet present, develops within second-stage cuticle.

Life cycle

Meloidogyne spp. are all root-endoparasites with a relatively simple life cycle and rapid rate of development and reproduction. The first-stage juvenile molts in the egg and hatches as a second-stage juvenile. This mobile stage migrates through the soil, invade root tips and moves intercellularly to their feeding site. Here they inject a few plant cells with gland cell secretions which develop into giant cells, if the plant is susceptible.

The second-stage juvenile becomes sedentary, feeds on the giant cells and increases in diameter to a characteristic flask-shaped stage with spiked tail. It moults three times without further feeding, there is no stylet visible in the third and fourth stage; male and female gonads start to develop.

After the last moult the adult stylet is formed, females continue feeding while growth is extremely rapid; they become saccate and start with egg-sac production and egg laying; large numbers of eggs (100-1000) are extruded into the gelatinous matrix. Adult males do not feed, they increase only in length and leave the root for mating. The length of the life cycle (egg to egg) is strongly influenced by temperature, but could be completed within 3 weeks; egg production however may continue 2 to 3 months (Bird, 1959; Trianthaphyllou & Hirschmann, 1960; Franklin, 1965; de Guiran & Ritter, 1979; Eisenback & Hirschmann, 1991).

Thesis outline

This study comprises a revision of the European root-knot nematodes, completed with a historical review of the genus *Meloidogyne* and a study on perineal pattern development.

Historical review

For a phytopathological significant nematode genus with nearly 150 years publication history, it is rather logical to find a relatively large number of published literature on the subject *Meloidogyne*. However only a few brief historical reviews are available on root-knot nematodes. While starting with the revision of the European root-knot nematodes several historical contradictions and unanswered questions were noticed.

Who described the perineal pattern for the first time? What is the correct author name and year of publication of *Meloidogyne exigua*: Goeldi, 1887; Goeldi, 1889; Göldi, 1892 or Goeldi, 1892? Who described the first detailed differences between the genera *Heterodera* and *Meloidogyne*? Why took it almost 100 years before root-knot nematodes were separated from cyst nematodes? If root-knot nematodes are so important, why are there only a few incomplete keys available for identification at present?

To answer these and other questions, the historical review (chapter 2) starts in the middle of the 19th century in Europe, a period when the relation between *Meloidogyne* and root-knot was discovered. Followed by an interesting but very confusing and relatively long period, in which there was no separation between *Heterodera* and *Meloidogyne*. After 1949, when the genus name *Meloidogyne* was finally re-introduced, the taxonomical papers and the number of species were booming.

This review includes a 'who described what and when' main part, focussed on species and taxonomical characteristics.

European root-knot nematodes

Two detailed monographic studies dealing with the whole genus *Meloidogyne* were published so far (Whitehead, 1968 & Jepson, 1987). While the former taxonomical study included 23 species, the latter about 51. Both monographs focus on the question 'how to identify all these different *Meloidogyne* species', and less on the specific species status.

More than 80 species of root-knot nematodes have been described between 1949 and 1998. Are all these species really nominal species? What about *Meloidogyne* species mixtures? And how many described, and probably undescribed, species occur in Europe?

Therefore a critical study (chapters 3-5) on the specific status of the European root-knot nematodes was initiated for the first time. Type specimens and additional live cultures were collected and compared with the original species descriptions. Live cultures were also used for isozyme electrophoresis as an additional characteristic. Three undescribed forms were detected in Europe, two extensive descriptions were added to this thesis. One recently described root-knot nematode was redescribed, as it was originally based on mixed species populations. Also information on species distribution and hosts was included, and identification keys to the European root-knot nematodes were prepared (Appendix II).

Perineal pattern

The perineal pattern or posterior pattern, is a unique but complex combination of several female characteristics. Although several authors consider the perineal pattern as highly variable, it is world-wide used as the most important character for *Meloidogyne* species identification.

Is the perineal pattern really such a highly variable character? What is the origin of this variability? What do we know about perineal pattern development? Is there a correlation between the second-stage juvenile tail and the final pattern?

Based on literature data and additional light and scanning electron microscopical observations on perineal patterns, a hypothesis for perineal pattern development was formulated (chapter 6). Additional observations on *Nacobbus aberrans* and *Cryphodera brinkmani* were added to extend the hypothesis to other plant parasitic nematodes genera with swollen females.

Literature

BALDWIN, J.G. (1992). Evolution of cyst and noncyst-forming Heteroderinae. *Ann. Rev. Phytopathol.* 30, 271-290.

BIRD, A.F. (1959). Development of the root-knot nematodes *Meloidogyne javanica* (Treub) and *Melodogyne hapla* Chitwood in tomato. *Nematologica* 4, 31-42.

EISENBACK, J.D. & HIRSCHMANN, H. (1991). Root-knot nematodes: *Meloidogyne* species and races. In: *Manual of agricultural nematology.* pp. 191-274. Ed. W.R. Nickle. New York, Marcel Dekker.

FRANKLIN, M.T. (1965). *Meloidogyne* root-knot eelworms. In: *Plant nematology, Technical Bulletin No. 7.* pp. 59-88. Ed. J.F. Southey. London, H.M.S.O.

GERAERT, E. (1997). Comparison of the head patterns in the Tylenchoidae (Nematoda). *Nematologica* 43, 283-294.

GUIRAN de, G. & RITTER, M. (1979). Life cycle of *Meloidogyne* species and factors influencing their development. In: *Root-knot nematodes* (Meloidogyne *species*) *Systematics, Biology and Control.* pp. 172-191. Eds. F. Lamberti & C.E. Taylor. London, Academic Press.

JEPSON, S.B. (1987). *Identification of root-knot nematodes* (Meloidogyne *species*). Wallingford, UK, C.A.B. International.

LUC, M., MAGGENTI, A.R. & FORTUNER, R. (1988). A reappraisal of Tylenchina (Nemata). The family Heteroderidae Filipjev & Schuurmans Stekhoven, 1941. *Rev. Nématol.* 11, 159-176.

MAGGENTI, A.R., LUC, M., RASKI, D.J., FORTUNER, R. & GERAERT, E. (1987). A reappraisal of Tylenchina (Nemata). 2. Classification of the suborder Tylenchina (Nemata: Diplogasteria). *Rev. Nématol.* 10, 135-142.

MAGGENTI, A.R. (1991). Nemata: higher classification. In: *Manual of agricultural nematology.* pp. 147-187. Ed. W.R. Nickle. New York, Marcel Dekker.

MÖHN, E. (1984). *System und Phylogenie der Lebewesen. Band 1. Physikalische, chemische und biologische Evolution. Prokaryonta. Eukaryonta (bis Ctenophora).* Schweizerbartsche Verlagsbuchhandlung, Stuttgart.

TRIANTAPHYLLOU, A.C. & HIRSCHMANN, H. (1960). Post-infection development of *Meloidogyne incognita* Chitwood, 1949 (Nematoda: Heteroderidae). *Ann. Inst. Phytopath. Benaki, N. S.* 3, 1-11.

WHITEHEAD, A.G. (1968). Taxonomy of *Meloidogyne* (Nematoda: Heteroderidae) with descriptions of four new species. *Trans. Zool. Soc. Lond.* 31, 263-401.

2

HISTORICAL NOTES ON THE GENUS

MELOIDOGYNE GÖLDI, 1892

'To obtain a definitive interpretation of an old name might
require more time than describing 20 new species in groups
not overburdened by old descriptions. In many cases, the
problems of old descriptions and old names are acute when
dealing with species from Europe or the Mediterranean
basin, simply because they have a large history of study.'

A. Minelli (1993).

Introduction

The taxonomical history of the genus *Meloidogyne* Göldi, 1892 can be divided into three
main parts. A period (1855-1878) in which a correlation was observed between root galls and
nematodes; followed by a relatively long confusing period (1879-1948) where root-knot
nematodes were included in the same genus as cyst nematodes (*Heterodera* Schmidt, 1871);
and finally the revival period, when root-knot nematodes were placed in a separate genus and
the number of described species and taxonomical papers were booming.

A few historical studies on root-knot nematodes are available. **Chitwood** (1949) briefly
described the taxonomical period between 1887 and 1949. **Gillard** (1961) was one of the first
who discussed the taxonomical and nomenclatural history from 1872 onwards. Followed by
two more comprehensive historical studies of **Whitehead** (1968) and **Franklin** (1979), both
starting in 1855.

The following review is more focused on the historical interesting period between 1855 and
1949, and chronologically presented. The period from 1949 onwards, is discussed briefly and
grouped in a number of species related subjects.

1855-1878

Although root-knot or galls on plants have been reported several times before 1855, the
cause was unknown or related with other organisms than nematodes. It was the Englishman
M.J. Berkeley (1855) who correlated, for the first time, galls on glasshouse cucumber roots
with nematodes. He observed in the roots an *'enormous development of the vascular tissue'*,
'certain cyst-like bodies filled with a multitude of minute eggs' and *'Vibrios or larvae'* free
and inside eggs. The cyst-like body was not a cyst but an adult *Meloidogyne* female. Berkeley
believed however that *'it is not conceivable that such a cyst could have been deposited by the
Vibrio itself'*, and described it as *'an irritation caused by the eggs'*. By denying the 'cyst' as

the adult stage he created a problem he was indirectly aware of *'The Vibrio was not seen except in a very young stage; any attempt, therefore, to ascertain the species, supposing it to be described, is hopeless'*.

In 1859 **Schacht** reported for the first time about a nematode disease on sugarbeet in Germany, in 1871 finally described by **Schmidt** as the cyst-forming sugarbeet nematode *Heterodera schachtii*.

Four years later, the Italian **Licopoly** (1875) described small nematodes within galls on the roots of *Sempervivum tectorum* L. Followed by the first report of root-knot on a field crop (coffee) in Brazil by **Jobert** (1878). He described different stages of nematode development, but incorrectly named the adult female a cyst.

Most textbooks on plant pathology start 'the early period' in 1853 with a publication of **Anton De Bary** on the association of a parasitic fungi *'Die Brand Pilze'* with rusts and smuts. The first report on root-knot nematode by **M.T. Berkeley** (1855), often overlooked, clearly belongs to one of the earliest contributions to plant pathology. **Heald** (1943) mentioned, in his introduction to Plant Pathology, about M.T. Berkeley *'and the contributions of M.T. Berkeley on British fungology and vegetable pathology (1846-1860) are also landmarks in the early modern era'*. It was also M.T. Berkeley who advised **Charles Darwin** with his experiments on seed dispersal in salt water, as published in the Origin of species (1859).

1879-1948

The first root-knot nematode was described in Europe by the Frenchman **Cornu** in 1879. He carefully studied root galls on different Leguminosae in the Loire valley, and observed in 1874 a root-knot nematode infection on *Onobrychis sativa* Lam. near Châteauneuf. Cornu illustrated galls, some of them with secondary roots, eggs with developing juveniles, a second-stage juvenile and a root section with females and protruding egg-masses. Although he described the females as cysts, to express the close relationship with *Heterodera schachtii*, he thought it was more related to *Anguillula radicicola* Greef, 1872 (= *Subanguina radicicola*), and described it as a new species *Anguillula marioni*.

The German **Müller** (1884) published an excellent and detailed morphological study on root-knot nematodes. He renamed *Anguillula radicicola* as *Heterodera radicicola* and synonymized *Anguillula marioni* with *H. radicicola*, this name was used for root-knot nematodes until 1932. Müller studied root-knot galls on several mono- and dicotyledons and because of the observed wide host range he concluded *'Die Gefahr, die dadurch unseren Kulturen droht, kann dabei nicht unterschätzt werden'*. He described in detail the process of gall induction and development, followed by morphological observations on stylet, dorsal pharyngeal gland orifice, metacorpus, gonads and spicules. It was Müller who observed and illustrated for the first time a perineal pattern and briefly described morphological differences between cyst and root-knot nematodes. One year later the German **Frank** (1885) published several new hosts of *H. radicicola* and introduced the idea to use hosts as 'Fangpflanzen' to reduce root-knot nematodes in the field.

The Dutchman **Treub** (1885) studied the origin of Sereh-disease (stunted plants) on sugarcane in Java, Indonesia. He detected a root-knot nematode and described it as a new species *Heterodera javanica*. The description however is very simple and only based on a few unreliable female and egg measurements. Not surprisingly **Beijerinck** (1887) and **van Breda de Haan** (1900) considered *Heterodera javanica* identical to *H. radicicola*.

In 1887 (published in 1892) **Göldi** described different diseases from coffee plants in Brazil, including a root-knot nematode. He named the nematode *Meloidogyne exigua* and described and illustrated it briefly. This publication however, was until 1949 ignored or unknown to most nematologist.

Neal (1889) published an interesting and extensive study on root-knot disease in the USA. In the introduction he mentioned about root-knot *'In 1878 I found the root-knot prevalent over Florida, and learned from old residents that as far back as 1805 it had been known, and from time immemorial had been dreaded as a foe to gardens and groves'*. He described the root-knot nematode as a new species *'I have provisionally named this worm Anguillula arenaria, but it may belong to a different genus'*, and added some measurements and illustrations. He described also the tail spike, characteristic for swollen juveniles, and the loss of it after the last moult, and probably illustrated a part of the female rectal glands. The diligent observations become more surprising if one realize that Neal was devoid of any published literature on this subject. He studied also the effect of disuse of land, soil types and fertilizers on root-knot nematode infection. On the effect of temperature he mentioned *'The question of temperature is no doubt one of great importance in determining the boundaries of this disease, perhaps more than food-plants or soils'*. In the same years **Atkinson** (1889) published a study on the life history of *Heterodera marioni* found on different plant species in the USA.

Stone & Smith (1898) published more details on different developmental stages of the root-knot nematode. They also illustrated the male coiled up in the juvenile cuticle.

In 1899, the Dutchman **van Breda de Haan**, reported about *H. radicicola* on tabacco in Deli (Indonesia). He tested some pesticides, studied the life history of the nematode, speculated for the first time about parthenogenetic mode of reproduction and observed a new structure for root-knot nematodes: the S-E duct and pore. He also discussed the function of the egg-sac *'in hun jongste stadium zijn de eieren wanneer zij juist buiten het moederdier zijn gekomen, zeer gevoelig voor uitdroging, hiertegen zal hun de slijmmassa beschermen en evenzoo tegengaan het binnendringen van parasieten van plantaardigen of dierlijke aard'*.

At the end of the 19th century, root-knot nematodes were known from all parts of the world, and it's polyphagous behaviour was recognized. No type material was deposited from the different 'species' described so far, even their type localities were vague. In the following 50 year the genus and species status of root-knot nematodes very slowly started to elucidate. The early 'taxonomical work' on the genus between 1910 and 1949 shifted from Europe to the USA, with a few exceptions.

Lavergne (1901a & b) reviewed root-knot records from South-America, studied root-knot in Chili and detected a nematode and described it as *Anguillula vialae*. He illustrated typical root-knot nematode galls from vineyards, some relatively small eggs, but a description of the root-knot nematode was not included.

The first woman who published on root-knot nematodes was **Kati Marcinowski** (1909). Her extensive and solid study on *'Parasitische und semiparasitisch an Pflanzen lebende Nematoden'*, includes 24 pages on cyst and root-knot nematodes. Marcinowski was the first who described in detail the morphological differences between female, male and juvenile cyst and root-knot nematodes. Although she recognized two different cyst species she believed that there was only one root-knot species, and synonymized the poorly described *M. exigua, H. javanica, A. arenaria, A. marioni* and *A. vialae* to *H. radicicola*. She hypothesized however *'Est ist möglich, daß sich bei genauerer hierauf gerichteter Untersuchung morphologisch charakterisierbare Varietäten von Heterodera radicicola werden auffinden lassen'*, she also

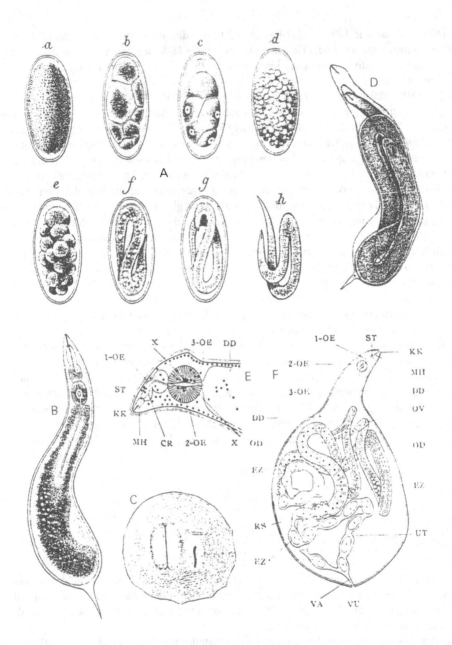

Early root-knot nematode illustrations. A: Egg development & B: Swollen second-stage juvenile (described as adult female), after Göldi, 1892; C: Perineal pattern & D: Male coiled in juvenile cuticle, after Müller, 1884; E: Female head & F: Female, after Nagakura, 1930.

mentioned briefly for the first time the possibility of host races, and listed 236 host plants. Two years later **Bessey** (1911) reported 480 hosts for *H. radicicola*.

Between 1911 and 1919 three studies on the control of *H. radicicola* were published by **Orton & Gilbert** (1912), **Bessey & Byars** (1915) and **Gilbert** (1917).

In 1919 **Kofoid & White** reported on a new nematode infection of man. They examined stools of about 140.000 soldiers in Texas (USA), and detected in 429 cases low numbers of an undescribed nematode ovum. They described it as '*This ovum is the largest ovum of intestinal worms encountered in human stools*' and '*No ova of like structure have been found by us on examination of feces of horses, mules, cattle, hogs, goats, or culture rats and mice from Texas*'. They observed in the eggs two highly refractive hyaline, bluish green globules flattened asymmetrically at the two poles of the embryo. These globules and the flattening were only observed in *Oxyuris* species, therefore they named the nematode *O. incognita*, as the adult stage was unknown. They were almost convinced that the unknown adult was not of plant nematode origin as '*Ova of plant nematodes presumably have no antidigestive ferments*'. **Hegner & Cort** (1921) confirmed the presence of *Oxyuris incognita* in human stools. **Sandground** (1922) observed differences in egg-sac formation, hatching and testis number within *H. radicicola* on two different hosts, possibly suggesting different species. In 1923 he published a paper titled '*Oxyuris incognita or Heterodera radicicola?*'. He proved that eggs of *H. radicicola* could be demonstrated in stools of man after the ingestion of parasitized roots, and observed the same polar globules and egg morphology as Kofoid & White (1919).

Based on morphological and parasitic behaviour differences between cyst and root-knot nematodes, **Cobb** (1924) proposed to place the latter in a new genus or subgenus *Caconema*. Although a rather logical proposal, it was not generally accepted.

The Japanese **Nagakura** (1930) studied '*Ueber den Bau und die Lebengeschichte der Heterodera radicicola*'. This relatively unknown paper was not even mentioned in the detailed reviews of Whitehead (1968) and Franklin (1979). Nagakura listed 12 morphological differences between *H. schachtii* and *H. radicicola*, studied intensively root-knot nematode morphology and observed the nematode life-cycle in sterile pure cultures. He described three juvenile stages (although he observed the first molt within the egg for the first time), three cuticle layers, stylet and pharynx, male and female gonad development and the six female rectal glands in detail.

Goodey (1932a & b) studied the biology of *Anguillulina radicicola* (Greef, 1872) and more important, reviewed the nomenclature of the root-gall nematodes. He concluded correctly that the gall inducing nematode *A. radicicola* belongs, to what we now name as *Subanguina radicicola*. The earliest available name for a root-knot nematode was *Anguillula marioni*, he therefore proposed *Heterodera marioni* as the correct name for root-knot nematodes. He disagreed with the proposal of Cobb to place root-knot nematodes in a separated genus, and synonymized *Caconema radicicola* with *Heterodera marioni*. The name *Heterodera marioni* was widely used for root-knot nematodes until 1949.

Tyler (1933a-b & 1938) studied respectively the asexual reproduction of the root-knot nematode in an aseptic root culture, the influence of temperature on nematode development and female egg-laying (up to 2000 eggs per female).

Shebakoff (1940) reported clear differences in host plant reaction with tomatoes and cotton. In a root-knot infested tomato field, cotton was not infected. In another root-knot cotton field however tomatoes were heavily infested. **Day & Tufts** (1940) observed inconsistent resistance with the same peach varieties between two nurseries, suggesting 'races' of *H. marioni*.

Christie & Cobb (1941) studied the life history of *H. marioni* and observed four moultings.

In 1941 **Filipjev & Schuurmans Stekhoven** published their 'Manual of agricultural helminthology', this important book included 8 cyst nematodes and *H. marioni* as the root-knot nematode. They were aware of the Göldi publication, and mentioned about the nomenclatorial action of Cobb *'There is however a still older generic name: Meloidogyne, which is considered to be a synonym of Heterodera. In the case in which such a distinction should be indispensable, it is the latter name and not that of Cobb that should be used'*.

It were **Christie & Albin** (1944) and **Christie & Havies** (1948) who finally proved the existence of 'races' within *H. marioni*, as discussed in detail by Whitehead (1949).

Almost one century after the first report of root-knot nematodes, they were still marked as a single polyphagous species within the genus *Heterodera*, despite the clearly documented differences between root-knot and cyst nematodes by **Marcinowski** (1909) and **Nagakura** (1930). These important papers were probably overlooked by most nematologist, even after the second World War.

1949-2000

In 1949 **Chitwood** published his 'revision of the genus *Meloidogyne* Goeldi, 1887', this relatively small paper thoroughly changed root-knot nematode taxonomy. Aware of the work of Christie *et al.* and the older literature on root-knot nematodes, he studied the morphology of the diverse kinds of described nematodes very well. On the morphology he mentioned *'Members of the genus Meloidogyne are extremely adaptable and their morphological characters show considerable variation. In fact we have not as yet seen two identical species. Nevertherless general pattern series and other structures show similarities, and progenies from individual females are relatively consistent both as to morphology and host range'*. He separated the root-knot nematodes from *Heterodera*, re-erected the genus *Meloidogyne*, and described a generic diagnosis for it. On species level Chitwood redescribed *M. arenaria*, *M. exigua*, *M. incognita* and *M. javanica*, and described *M. hapla* and a variety *M. incognita* var. *acrita*. The species concept he introduced was based on perineal pattern morphology added with stylet knob shape, stylet and DGO length differences.

The nomenclatorial action of Chitwood was correctly criticized by **Gillard**, 1961 and **Whitehead**, 1968 (see also chapter 5). Beside this nomenclatorial question, the Chitwood paper was undoubtly the starting point for root-knot nematode taxonomy.

Chitwood described Goeldi, 1887 as the author name and year of publication for the genus *Meloidogyne*, although it was published in **1892** according to **Fortuner**, 1984 and **Wouts & Sher**, 1971. Also other incorrect data as 1889 and 1894 have been published. Most publications after 1949 on *Meloidogyne* mentioned Goeldi as the *M. exigua* author, but is this correct? In the original publication, Goeldi is named as **Göldi**, with a diacritic mark. ICZN article 27 is clear on the use of diacritic marks *'No diacritic mark, apostrophe, or diaeresis is to be used in a scientific name regulated by the Code'*. However, article 51a mentioned about the author name *'The name of the author does not form part of the name of a taxon...'*. Therefore the original name **Göldi** is the correct author name of *M. exigua*. Even in the extensive 'Bibliography and host list' on the genus *Heterodera* (**Anon.**, 1931), it was already listed as Göldi, 1892.

Species descriptions

Between 1949 and the year 1998 more then eighty nominal *Meloidogyne* species have been described (Appendix I), especially from North and South-America, Africa, China and Europe. More than 50% of the species have been described during the last twenty years. The greater part of these species were detected in agricultural areas and described from food crops. Not surprisingly we may expect more new forms in the future, particularly described from natural areas.

The first named species after Chitwood, 1949 were morphologically very distinct (in particular the perineal pattern), and based on light microscopical differences in a limited number of female, male and second-stage juvenile characters, added with some host plant differences (i.e. *M. brevicauda* Loos, 1953; *M. acronea* Coetzee, 1956; *M. africana* Whitehead, 1960; *M. coffeicola* Lordello & Zamith, 1960 and *M. ovalis* Riffle, 1963).

In 1952 **Sasser** introduced a method for identification of the 'Chitwood species' by host reaction. This test was often included in later descriptions to proof that a new species showed a 'unique' host reaction (see also **Sasser**, 1979). This test is now known as the 'North Carolina differential host test' and also used to detect host races within the 'Chitwood species' (**Sasser & Carter**, 1985).

Sledge & Golden (1964) separated a new root-knot nematode with an elevated perineal pattern and a thicker cuticle into a new genus *Hypsoperine* (*H. graminis*). They also renamed *Meloidogyne acronea* as *Hypsoperine acronea*. Between 1964 and 1968 the total number of *Hypsoperine* species raised to six. **Whitehead** (1968) synonymized *Hypsoperine* with *Meloidogyne*, as the elevated pattern and the thicker cuticle are rather variable characters and recorded in several other *Meloidogyne* species.

Spaull (1977) introduced for the first time scanning electron microscopical (SEM) images in a root-knot nematode description (*M. propora*). After this publication, SEM images have become an essential part in most root-knot nematode descriptions.

Several comparative SEM and LM studies on morphology in the seventies and eighties, increased the number of useful descriptive characters for eggs, males, females and second-stage juveniles up to 140 (see **Jepson**, 1987 for a review).

Historically, research on cytogenetics, isozymes and DNA of root-knot nematodes is recent. Therefore only for a limited number of species, described in the last twenty years, one may find these useful characteristics in species descriptions.

A critical look at Appendix I shows a relatively low number of species synonyms, mainly restricted to the 'Chitwood species'. Also the number of *species inquirendae* is very low, especially if one compares it with other well studied plant parasitic nematode genera, like *Pratylenchus* with 46 species, 20 synonyms and 28 *species inquirendae* (**Loof**, 1991). It illustrates the present state of *Meloidogyne* taxonomy, mainly descriptive and only a few detailed comparative studies available. Even these studies are not focused on the species status, but on the question 'how to identify?'.

Root-knot nematode descriptions have evolved during the last fifty years from relatively simple to complex pieces of 'art work'. Partly because of this complexity, it is for a diagnostic laboratory at present impossible to identify al these described species. There is simply not a recent up to date and reliable identification key available for all 'species'. Another reason is the 'classical taxonomical nightmare': incomplete descriptions and poorly preserved or non-available type material. There is clearly a need for a world-wide revision of the genus *Meloidogyne*.

European species

The following, chronologically listed, *Meloidogyne* species have been described [*] or detected in Europe:

-*M. marioni* (Cornu, 1879) Chitwood & Oteifa, 1952 (*species inquirenda*).*
-*M. javanica* (Treub, 1885) Chitwood, 1949.
-*M. arenaria* (Neal, 1889) Chitwood, 1949.
-*M. exigua* (Göldi, 1892) Chitwood, 1949.
-*M. incognita* (Kofoid & White, 1919) Chitwood, 1949.
-*M. hapla* Chitwood, 1949.
-*M. artiellia* Franklin, 1961.*
-*M. graminis* (Sledge & Golden, 1964) Whitehead, 1968.
-*M. naasi* Franklin, 1965.*
-*M. kirjanovae* Terenteva, 1965.*
-*M. ardenensis* Santos, 1968.*
-*M. deconincki* Elmiligy, 1968.*
-*M. litoralis* Elmiligy, 1968.*
-*M. megriensis* (Poghossian, 1971) Esser, Perry & Taylor, 1976.*
-*M. chitwoodi* Golden, O'Bannon, Santo & Finley, 1980.
-*M. kralli* Jepson, 1983.*
-*M. hispanica* Hirschmann, 1986.*
-*M. maritima* (Jepson, 1987) Karssen, van Aelst & Cook, 1998.*
-*M. lusitanica* Abrantes & Santos, 1991.*
-*M. fallax* Karssen, 1996.*
-*M. duytsi* Karssen, van Aelst & van der Putten, 1998.*

Franklin described the first root-knot nematode from Europe in 1961. She published an excellent description, followed in 1965 by her second detailed *Meloidogyne* description. With these publications the taxonomical work on the genus turned back to Europe. From the 20 root-knot nematodes detected in Europe, 8 species were described by woman (**Franklin, Terenteva, Santos, Jepson, Hirschmann** and **Abrantes**). Although the reason is unknown to me, it is a high number if compared to other plant parasitic nematode genera. All these species descriptions have a common characteristic: very detailed and well illustrated.

Most species were described from agricultural areas, except for *M. ardenensis, M. deconincki, M. litoralis, M. kralli, M. maritima* and *M. duytsi*.

Taxonomical and related studies

Between 1950 and 1960 several contradicting studies were published on the taxonomical value of the perineal pattern (see chapter 6).

Until 1968 *Meloidogyne* taxonomy was mainly descriptive, with a few exceptions. **Taylor** *et al.*, 1955 studied perineal pattern variation of 8 species, and included a key. **Goffart** (1957) and **Franklin** (1957) published both brief reviews on 11 described species.

Gillard (1961) studied the biology, distribution and control of root-knot nematodes. This thesis includes also a key towards 11 species, a brief historical review and the first detailed critical remarks on the species status of the 'Chitwood species'.

Whitehead (1968) published the first extensive comparative monography on 23 known species. He studied in detail female, male and second-stage juvenile morphology, and proposed two keys (including the first key for J2's).

Franklin (1972) reviewed 32 root-knot nematodes and focussed in this paper on second-stage juvenile and perineal pattern morphology.

In 1975 an **International *Meloidogyne* Project** (IMP) started, it included about 100 scientists from 43 nations. The main goal was to assist developing nations in increasing yields of economic field crops, by promoting knowledge about root-knot nematodes and to protect crops from the damaging effects of root-knot. The IMP continued for 8 years, with an enormous spin-off in *Meloidogyne* papers on biology, taxonomy, ecology, management and techniques. Also 3 books were published 'Root-knot nematodes (*Meloidogyne* species)' edited by **Lamberti & Taylor** (1979) and 'An advanced treatise on *Meloidogyne*' volume I & II, both edited by **Sasser & Carter** (1985). Although it is impossible to discuss all IMP papers and related publications in detail, two subjects however need attention: cytogenetics and enzyme phenotypes.

Triantaphyllou (1979 & 1985) studied about 600 populations (24 species) of *Meloidogyne* cytogenetically, in particular the method of reproduction and chromosome number. The majority of the studied species reproduce by parthenogenesis, only a few rare species reproduce by amphimixis. He distinguished two groups within the parthenogenetic species, one diploid group with a chromosome number of n = 18, reproducing (facultative) meiotic parthenogenesis; the other polyploid group ranging in chromosome number (2n) from 30 to 56, reproducing (obligatory) mitotic parthenogenesis. The latter includes the most successful *Meloidogyne* species as *M. arenaria*, *M. incognita*, *M. javanica* and some *M. hapla* populations.

Esbenshade & Triantaphyllou (1985a,b & 1987) selected 300 populations, of the 600 which had been studied cytologically, and studied several enzymes. They also listed most of the proteins and enzymes detected in *Meloidogyne* species (1985a). Four isozymes (esterase, malate dehydrogenase, superoxide dismutase and glutamate-oxaloacetate transaminase) proved to be very useful for routine identification, because the product of one female is sufficient for detection and the variability of these enzymes within a species is very low.

Jepson published in 1987 her book on the genus *Meloidogyne*, for the morphological identification of 51 species based on female, male and second-stage juvenile characters. She included useful lattice keys and information on host, distribution, symptoms and techniques. This book and the IPM books refer to detailed morphological studies on root-knot nematodes made by **Hirschmann**, **Eisenback** and **Jepson**.

In the last decade several articles were published on **DNA** level. Most of them proposing different methods for identification of a limited number of root-knot nematodes. Undoubtedly more comprehensive studies on this subject will be published in the nearby future.

In 1990 and 1993 two books on plant parasitic nematodes in subtropical and tropical (Ed. **Luc** *et al.*) respectively temperate agriculture (Ed. **Evans** *et al.*), appeared. Both include a lot of information on the genus *Meloidogyne*, in particular on host plants.

The successor of 'A manual of agricultural helminthology' (1941), the 'Manual of agricultural nematology' by **Nickle**, was published in 1991. And includes an extensive chapter on nine root-knot nematodes, written by **Eisenback & Hirschmann**.

Recently the first CD-ROM concerned with one plant parasitic nematode genus was published (**Eisenback**, 1998). This root-knot nematode taxonomic database includes the most important articles and book chapters on cytology, distribution, identification, molecular biology, morphology and techniques (Karssen & Wiesenekker, 1998).

Literature

ANON. (1931). *The root-infesting eelworms of the genus* Heterodera, *a Bibliography and Host List*. Imp. Bureau of Agricultural parasitology, St. Albans, UK.

ATKINSON, G.F. (1889). A preliminary report upon the life history and metamophoses of a root-gall nematode, *Heterodera marioni* (Greef) Müller, and the injuries caused by it upon the roots of various plants. *Sci. Contr. Agric. Exp. Stn. Alabama Polyt. Inst.* 1, 177-226.

BEIJERINCK, M.W. (1887). The Gardenia root-disease. *Garderners Chronicle* 1, 488-489.

BERKELEY, M.J. (1855). 'Vibrio forming cysts on the roots of cucumbers'. *Gardeners' Chronicle* April 7th, 220.

BESSEY, E.A. (1911). Root-knot and its control. *Bull. U.S. Dept. Agric.* 217.

BESSEY, E.A. & BRYARS, L.P. (1915). The control of Root-Knot. *U.S. Dept. Agric. Farmer's Bull.* 648, 1-19.

BREDA DE HAAN, J. VAN. (1899). Levensgeschiedenis en Bestrijding van het Tabaks-aaltje (*Heterodera radicicola*) in Deli. *Meded. Pl.Tuin, Batavia* 35.

BREDA DE HAAN, J. VAN. (1900). Die Lebensgeschichte des Tabaksälchens (*Heterodera radicicola*) und seine Bekämpfung in Deli. *Bull. de l'inst. bot. de Buitenzorg* 4.

CHITWOOD, B.G. (1949). Root-knot nematodes, part I. A revision of the genus *Meloidogyne* Goeldi, 1887. *Proc. Helminth. Soc. Wash.* 16, 90-104.

CHRISTIE, J.R. & COBB, G.S. (1941). Notes on the life history of the root-knot nematode, *Heterodera marioni*. *Proc. helminth. Soc. Wash.* 8, 23-26.

CHRISTIE, J.R. & ALBIN, F.E. (1944). Host-parasite relationships of the root-knot nematode. I. The question of races. *Proc. helminth. Soc. Wash.* 11, 31-37.

CHRISTIE, J.R. & HAVIS, L. (1948). Relative susceptibility of certain peach stocks to races of the root-knot nematode. *Plant Dis. Reptr.* 32, 510-514.

COBB, N.A. (1924). The amphids of *Caconema* (*nom. nov.*) and other nemas. *J. Parasit.* 11, 118-120.

CORNU, M. (1879). Etudes sur le *Phylloxera vastatrix*. *Mém. Divers Sav. Acad. Sc. Institut France* 26, 163-175, 328, 339-341.

DAY, L.H. & TUFTS, W.P. (1940). Further notes on nematode-resistant rootstocks for deciduous fruit trees. *Proc. Am. Soc. hort. Sci.* 37, 327-329.

EISENBACK, J.D. & HIRSCHMANN, H. (1991). Root-knot nematodes: *Meloidogyne* species and races. In: *Manual of agricultural nematology*. pp. 191-274. Ed. W.R. Nickle. New York, Marcel Dekker, inc.

EISENBACK, J.D. (1998). *Root-knot nematode taxonomic database*. CD-ROM, Wallingford, UK, C.A.B. International.

ESBENSHADE, P.R. & TRIANTAPHYLLOU, A.C. (1985a). Identification of major *Meloidogyne* species employing enzyme phenotypes as differentiating characters. In: *An advanced treatise on* Meloidogyne. *Volume I, Biology and Control*. pp. 135-140. Ed. J.N. Sasser & C.C. Carter. Raleigh, USA, North Carolina State University Graphics.

ESBENSHADE, P.R. & TRIANTAPHYLLOU, A.C. (1985b). Use of enzyme phenotypes for identification of *Meloidogyne* species. *J. Nematol.* 17, 6-20.

EVANS, K., TRUDGILL, D.L. & WEBSTER, J.M. (1993). *Plant parasitic nematodes in temperate agriculture*. Wallingford, UK, C.A.B. International.

ESBENSHADE, P.R. & TRIANTAPHYLLOU, A.C. (1987). Enzymetic relationships and evolution in the genus *Meloidogyne* (Nematoda: Tylenchida). *J. Nematol.* 19, 8-18.

FILIPJEV, I.N. & SCHUURMANS STEKHOVEN, J.H. (1941). *A manual of agricultural helminthology*. Leiden, E.J. Brill.

FORTUNER, R. (1984). List and status of the genera and families of plant-parasitic nematodes. *Helminth. Abstracts* 53, 87-133.

FRANK, A.B. (1885). Über das Wurzelälchen und die durch dasselbe verursachten Beschädigungen der Pflanzen. *Landw. Jahrb.* 14, 149-176.

FRANKLIN, M.T. (1957). Review of the genus *Meloidogyne*. *Nematologica* 2, 387-397.

FRANKLIN, M.T. (1961). A British root-knot nematode, *Meloidogyne artiellia* n. sp. *J. Helminth.*, R.T. Leiper Suppl. 85- 92.

FRANKLIN, M.T. (1965). A root-knot nematode, *Meloidogyne naasi* n. sp., on field crops in England and Wales. *Nematologica* 11, 79-86.

FRANLIN, M.T. (1972). The present position in the Systematics of *Meloidogyne*. *OEPP/EPPO bull.* 6, 5-15.

FRANKLIN, M.T. (1979). Taxonomy of the genus *Meloidogyne*. In: *Root-knot nematodes* (Meloidogyne *species), Systematics, Biology and Control*. pp. 37-54. Ed. F. Lamberti & C.E. Taylor. London, Academic Press.

GILLARD, A. (1961). Onderzoekingen omtrent de biologie, de verspreiding en de bestrijding van wortelknobbelaaltjes (*Meloidogyne* spp.). *Meded. LandbHoogesch., Gent* 26, 515-646.

GÖLDI, E.A. (1892). Relatoria sôbre a molestia do cafeiro na provincia da Rio de Janeiro. *Archos Mus. nac., Rio de Janeiro* 8, 7-112.

GOFFART, H. (1957). Bemerkungen zu einigen arten der gattung *Meloidogyne*. *Nematologica* 2, 177-184.

GOODEY, T. (1932a). On the nomenclature of the root-gall nematode. *J. Helminth.* 10: 21-28.

GOODEY, T. (1932b). The Genus *Anguillulina* Gerv & v. Ben., 1859, vel *Tylenchus* Bastian, 1865. *J. Helminth.* 10: 75-180.

HEALD, F.D. (1943). *Introduction to Plant Pathology.* New York, McGraw-Hill Book Company, inc.

HEGNER, R.W. & CORT, W.W. (1921). *Diagnosis of protozoa and worms.* Baltimore, USA.

JEPSON, S.B. (1987). *Identification of root-knot nematodes* (Meloidogyne *species*). Wallingford, UK, C.A.B. International.

JOBERT, C. (1878). Sur une maladie du cafeier observée au Bresil. *C.r. hebd. Séanc. Acad. Sci., Paris* 87, 941-943.

KARSSEN, G. & WIESENEKKER, E. (1998). Review on 'Root-knot nematode taxonomic database'. CD-ROM, Ed. J.D. Eisenback. *Nematologica* 44, 789-790.

KOFOID, C.A. & WHITE, W.A. (1919). A new nematode infection of man. *J. Am. med. Ass.* 72, 567-569.

LAVERGNE, G. (1901a). La anguilula en Sud-America. *Revista Chilena de Historia Naturel, Valpariosa* 4, 85-91.

LAVERGNE, G. (1901b). L'anguillula du Chili (*Anguillula vialae*). *Rev. Viticulture* 16, 445-451.

LICOPOLY, G. (1875). Sopra alcuni tubercoli radicellari continente anguillole. *Rc. Accad. Sci. fiz. Napoli* 14, 41-42.

LOOF, P.A.A. (1991). The family Pratylenchidae Thorne, 1949. In: *Manual of Agricultural Nematology.* pp. 363-421. Ed. W.R. Nickle. New York, Marcel Dekker, Inc.

LUC, M., SIKORA, R.A. & BRIDGE, J. (1990). *Plant parasitic nematodes in subtropical and tropical agriculture.* Wallingford, UK, C.A.B. International.

MARCINOWSKI, K. (1909). Parasitisch und semiparasitisch an Pflanzen lebende Nematoden. *Arbeiten aus der Kaiserlichen Biologischen Anstalt für Land- und Forstwirtschaft* 7 (1), 1-192.

MINELLI, A. (1993). *Biological systematics, the state of the art.* London, UK, Chapman & Hall.

MÜLLER, C. (1884). Mittheilungen über die unseren Kulturpflanzen schädlichen, das Geschlecht *Heterodera* bildenden Würmer. *Landw. Jahrb.* 13, 1-42.

NAGAKURA, K. (1930). Ueber den Bau und die Lebensgeschichte der *Heterodera radicicola* (Greef) Müller. *Japanese J. Zoology* 3, 95-160.

NEAL, J.C. (1889). The root-knot disease of the peach, orange and other plants in Florida, due to the work of the *Anguillula. Bull. U.S. Bur. Ent.* 20, 1-31.

ORTON, W.A. & GILBERT, W.W. (1912). Control of cotton wilt and root knot. *Circ. U.S. Bur. Pl. Ind.*, 92.

SANDGROUND, J. (1922). A study of the life-history and methods of control of the root gall nematode *Heterodera radicicola. South African Journ. Sci.* 18, 399-418.

SANDGROUND, J. (1923). "*Oxyuris incognita*" or *Heterodera radicicola*? *J. Parasit.* 10, 92-94.

SASSER, J.N. (1952). Identification of root-knot nematodes (*Meloidogyne* spp.) by host reaction. *Plant Dis. Reptr.* 36, 84-86.

SASSER, J.N. (1979). Pathogenicity, host ranges and variability in *Meloidogyne* species. In: *Root-knot nematodes* (Meloidogyne *species*), *Systematics, Biology and Control.* pp. 257-268. Ed. F. Lamberti & C.E. Taylor. London, Academic Press.

SASSER, J.N. & CARTER, C.C. (1985). Overview of the International *Meloidogyne* Project. In: *An advanced Treatise on* Meloidogyne. *Volume I, Biology and Control.* pp. 19-24. Ed. J.N. Sasser & C.E. Taylor. Raleigh, USA, North Carolina State University Graphics.

SCHACHT, H. (1859). Über einige Feinde und Krankheiten der Zuckerrübe. *Zeitschr. Rübenzucker-Industrie* 9, 239-250.

SCHMIDT, A. (1871). Über den Rübennematoden (*Heterodera schachtii*). *Zeitschr. Rübenzucker-Industrie* 21, 1-19.

SHERBAKOFF, C.D. (1940). Recent field observations on tomato and cotton root-knot nematodes. *Plant Dis. Reptr. Suppl.* 124, 146.

SLEDGE, E.B. & GOLDEN, A.M. (1964). *Hypsoperine graminis* (Nematoda: Heteroderidae) a new genus and species of plant parasitic nematode. *Proc. Helminth. Soc. Wash.* 31, 83-88.

SPAULL, V.W. (1977). *Meloidogyne propora* n. sp. (Nematoda: Meloidogynidae) from Aldabra atoll, Western Indian ocean, with a note on *M. javanica* (Treub). *Nematologica* 23, 177-186.

STONE, G.H. & SMITH, R. (1898). Nematode Worms. *Hatch. Exp. Stat. Mass. Bull.* 55, 1-67.

TAYLOR, A.L., DROPKIN, V.H. & MARTIN, G.C. (1955). Perineal patterns of root-knot nematodes. *Phytopathology* 45, 26-34.

TYLER, J. (1933a). Reproduction without males in aseptic root cultures of the root-knot nematodes. *Hilgardia* 7, 373-388.

TYLER, J. (1933b). Development of the root-knot nematode as affected by temperature. *Hilgardia* 7, 389-415.

TYLER, J. (1938). Egg output of the root-knot nematode. *Proc. Helminthol. Soc.* 5, 49-54.

TREUB, M. (1885). Onderzoekingen over sereh-ziek suikerriet gedaan in 'slands plantentuin te Buitenzorg. *Meded. Pl.Tuin, Batavia* 2.

TRIANTAPHYLLOU, A.C. (1979). Cytogenetics of root-knot nematodes. In: *Root-knot nematodes* (Meloidogyne *species*), *Systematics, Biology and Control*. pp. 85-109. Ed. F. Lamberti & C.E. Taylor. London, Academic Press.

TRIANTAPHYLLOU, A.C. (1985). Cytogenetics, cytotaxonomy and phylogeny of root-knot nematodes. In: *An advanced treatise on* Meloidogyne. *Volume I, Biology and Control*. pp. 113-126. Ed. J.N. Sasser & C.C. Carter. Raleigh, USA, North Carolina State University Graphics.

WHITEHEAD, A.G. (1968). Taxonomy of *Meloidogyne* (Nematoda: Heteroderidae) with descriptions of four new species. *Trans. zool. Soc. Lond.* 31, 263-401.

WOUTS, W.M. & SHER, S.A. (1971). The genera of the subfamily Heteroderinae (Nematoda: Tylenchoidae) with a description of two new genera. *J. Nematol.* 3, 129-144.

3

REVISION OF THE

EUROPEAN ROOT-KNOT NEMATODES

I. ON DICOTYLEDONS

'In this group there are three European species with overlapping hosts and very similar morphology. Only *M. ardenensis* has been closely studied: populations of *M. deconincki* and *M. litoralis* have not been kept anywhere in culture and the type locality of *M. deconincki* no longer exists.'

S.B. Jepson (1987).

Introduction

Root-knot nematodes are a genus of sedentary plant-parasitic nematodes with a world-wide distribution. Up to 1998 more than 80 nominal species have been described. The genus is easy to recognize but identification to species level is difficult due to relatively small inter-specific morphological variation (Jepson, 1987).

Brief taxonomical studies dealing with a part or the whole genus *Meloidogyne* were published by Chitwood (1949), Goffart (1957), Franklin (1957, 1972, 1979), Esser *et al.* (1976), Hewlett & Tarjan (1983), Hirschmann (1985), Kleynhans (1989) and Eisenback & Hirschmann (1991). Whitehead (1968) and Jepson (1987) published detailed monographs.

Twenty root-knot nematode species have been detected in Europe so far (chapter 2), thirteen of them having been described from an European type locality. Recently it has become clear that at least one species description was based on mixed species populations (Karssen *et al.*, 1998). Probably other species are also doubtful. A taxonomic study has been initiated in which the specific status of all the species recorded in Europe, including three undescribed forms, is critically assessed for the first time.

For this purpose the species were grouped into three field host groups. In this chapter the *Meloidogyne* species mainly parasitizing dicotyledonous field hosts are discussed.

Material and Methods

Type specimens of *M. ardenensis*, *M. deconincki*, *M. exigua*, *M. hapla*, *M. litoralis* and *M. lusitanica* from different nematode collections were used for comparison with the original species descriptions and additional taxonomical studies. Measurements and photographs were taken with a light microscope using differential interference contrast, drawings were made with a drawing tube.

Subsequently, live cultures of the nematodes were obtained, where possible, from the type localities. Specimens from these populations were examined by light microscopy (data not presented), prepared as described in detail in the following chapter (for practical reason), and used for isozyme electrophoresis (Esbenshade & Triantaphyllou, 1985 & 1987; Karssen et al., 1995).

The isozymes malate dehydrogenase (Mdh; EC 1.1.1.37) and esterase (Est; EC 3.1.1.1) were tested and used as an additional taxonomical character. In all the presented isozyme patterns, M. javanica is used as a stable reference and placed in the two middle positions of the gel. The presention of enzyme patterns used for all studied populations is the product of one young egg-laying female per lane. At least twenty individuals per population were studied.

The species descriptions are presented in a relatively simple format, focused on striking morphological and morphometrical characteristics of the type material, together with schematic drawings and additional species information.

Species descriptions

Meloidogyne ardenensis Santos, 1968

(Fig. 1-3)

Meloidogyne ardenensis

Santos (1968); *Nematologica* 13: 593-598.
Sturhan (1976); *Nachrichtenbl. Deut. Pflanzenschutzd.* 28: 113-117.
Thomas & Brown (1981); *Pl. Path.* 30: 147-151.
Stephan & Trudgill (1982); *Revue Nématol.* 5: 281-284.
Jepson (1987); *Identification of root-knot nematodes* (Meloidogyne *species*). Wallingford, UK: CAB International. 265 pp.

Measurements

(in glycerine) see Table I

Female

Body pear shaped, relatively small with long curved neck and thickening behind metacorpus. Perineal pattern located on a slight posterior protuberance. Long stylet, cone slightly curved dorsally, knobs ovoid and backwardly sloping. S-E pore less than one stylet length behind head end. DGO to stylet knobs relatively long. Perineal pattern oval to angular shaped with coarse striae and high dorsal arch, lateral field obscure.

Male

Relatively rare. Head slightly set off from body and relatively high. Head cap distinct with elevated labial disc, lateral lips present. Head region without transverse incisures. Long stylet, knobs backwardly sloping and ovoid shaped. DGO to stylet knobs relatively long. Lateral field with four incisures, outer bands areolated, fifth incomplete incisure often present.

Second-stage juvenile

Body moderately long and relatively stout. Hemizonid posterior to the S-E pore. Pharyngeal gland lobe well developed, ventrally overlapping of intestine sometimes difficult to observe. Tail conical-shaped with relatively short tail and hyaline tail part. Rectum rarely inflated. Anterior region of hyaline tail part with ventral indent in epidermis. Hyaline tail part with cuticular constriction, ending in broadly rounded terminus.

Hosts

Detected mainly on trees, shrubs and some dicotyledonous herbs (Jepson, 1987). On woody plants it induces very small galls. Incidentally it is reported on the Iridaceae *Hosta* sp. (Richter, 1981). Although *M. ardenensis* has not been recorded on Poaceae field hosts so far, Thomas & Brown (1981) detected some experimental Poaceae hosts with a Scottish population.
I have found this species recently in the meadows of the river Rhine near Wageningen, the Netherlands, on *Crataegus monogyna* L. and in the forest 'Leuvenumse bos' near Harderwijk, the Netherlands, and the coastal dunes near Oostvoorne, the Netherlands, on another Rosaceae *Rubus fruticosis*. Both are new hosts.

Distribution

M. ardenensis is reported from the Netherlands, Germany, UK (Jepson, 1987), Poland (Brzeski, 1996), Russia (Chizhov *et al.*, 1986), Belgium and France (see *M. deconincki* and *M. litoralis*, this chapter). It is not reported outside Europe so far.

Type locality & host

Woodland near Ranby UK, described from *Vinca minor* L., at the type locality it was also found on *Ligustrum vulgare* L. and *Sambucus* sp.

Type material

Paratypes deposited at Universidade de Coimbra, Coimbra Portugal and Rothamsted Experimental Station, Harpenden UK, were studied. Additionally the slides used for the book of Jepson (1987) were studied; they originated from *Vinca minor* L. near Bristol UK.

Etymology

The species name refers to Arden, an old colloquial name for Harpenden.

Remarks

The studied slides and additional populations are in agreement with the description, except for a slightly longer observed DGO to stylet knobs distance in males and females. The observed ventral indent in the anterior region of second-stage juvenile tails, was illustrated but not specified in the original description (Fig. 3I). Although the origin and functional meaning of this indent is unknown, it is only observed in *M. ardenensis* J2's so far and therefore a useful diagnostic character.

Different populations from the Netherlands, Belgium and Germany revealed a N1a type malate dehydrogenase pattern and a faint multiple banding esterase pattern (Fig. 3).

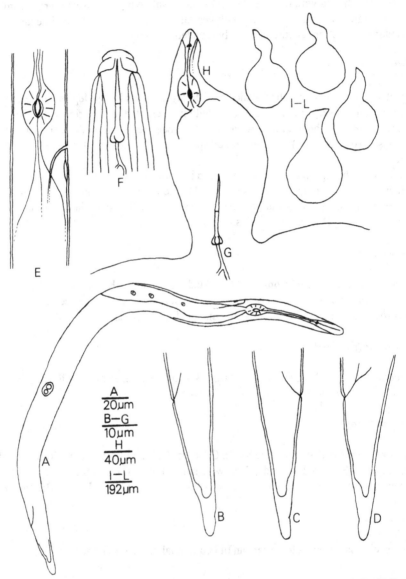

Fig. 1. *Meloidogyne ardenensis*. A-E: Second-stage juvenile (lateral). A: Body; B-D: Tails; E: Metacorpus region; F: Male head end (lateral); G-L: Female (lateral). G: Stylet; H: Anterior end; I-L: Body shapes (specimens from Bristol UK).

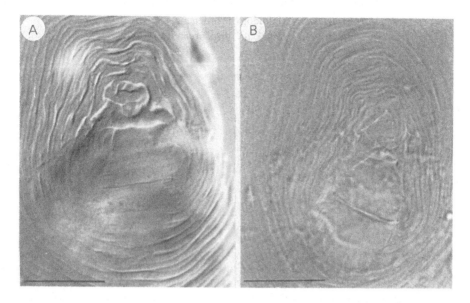

Fig. 2. *Meloidogyne ardenensis.* A & B: Female perineal patterns. Bar= 25 μm.

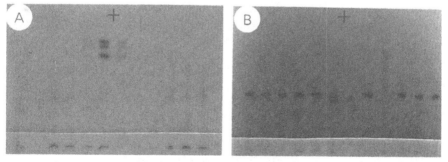

Fig. 3. *Meloidogyne ardenensis.* Esterase (A) and malate dehydrogenase patterns (B) of populations from the Netherlands (Oostvoorne and Wageningen), Belgium (Gent) and Germany (Köln); reference: *M. javanica* (China) in the two middle positions of the gel.

Table I

Morphometrics of Meloidogyne ardenensis *Santos, 1968*
[*mean ± SD (range); all measurements in µm*]

Character	J2	Males	Females
N	11	10	9
Body length	407±27.5	1609±272	-
	(365-451)	(1062-1939)	
Greatest body diameter	22.5±3.4	39.5±3.8	-
	(18.3-27.2)	(32.9-44.2)	
Body diam. at stylet knobs	-	17.6±1.2	-
		(15.8-19.0)	
„ „ „ S-E pore	16.1±1.8	27.2±1.7	-
	(13.3-18.3)	(24.7-29.1)	
„ „ „ anus	11.9±1.4	-	-
	(10.1-14.5)		
Stylet length	12.4±0.5	22.5±0.6	18.2±0.7
	(12.0-13.2)	(21.5-23.5)	(17.1-19.0)
DGO	3.5±0.3	5.4±0.3	5.6±0.5
	(3.2-3.8)	(5.1-5.7)	(5.1-6.3)
S-E pore to anterior end	-	-	11.1±2.4
			(7.6-14.5)
Anterior end to metacorpus	-	74±5.0	76±12.2
		(64-82)	(63-95)
Metacorpus length	-	-	38.5±3.0
			(34.8-45.5)
„ diameter	-	-	30.2±1.4
			(28.4-32.9)
Tail length	39.7±1.8	-	
	(36.7-41.7)		
Tail terminus length	11.6±0.6	-	-
	(10.7-12.6)		
Anus-primordium	102±7.8	-	
	(95-119)		
Spicule	-	36.5±1.0	-
		(35.4-37.9)	
Gubernaculum	-	9.8±0.4	-
		(9.5-10.7)	
a	18.5±2.9	40.7±4.7	-
	(15.3-24.3)	(31.6-46.5)	
b"	6.4±0.5	-	-
	(5.5-7.0)		
c	10.3±0.7	-	-
	(9.6-11.2)		
c'	3.4±0.3	-	-
	(2.9-3.9)		
T	-	52±7.2	-
		(38-60)	
(S-E pore/L)x 100	17.8±1.0	9.1±1.4	-
	(16.9-19.8)	(7.8-12.4)	

(b"= body length: distance from anterior end to centre of medium bulb)

Meloidogyne deconincki Elmiligy, 1968

[=Syn. with *M. ardenensis*]

(Fig. 4-6)

Meloidogyne deconincki

Elmiligy (1968); *Nematologica* 14: 577-590.
 Jepson (1987); *Identification of root-knot nematodes* (Meloidogyne *species*). Wallingford,
 UK: CAB International. 265 pp.
 Bongers (1988); *De nematoden van Nederland.* Utrecht: KNNV. 408 pp.

Measurements

(in glycerine) in Table II

Female

Holotype. Pear shaped body, relatively long curved neck and thickening behind metacorpus.
Long stylet, cone slightly curved dorsally, knobs ovoid and backwardly sloping. S-E pore less
than one stylet length behind head end. Distance DGO to stylet knobs relatively long. Perineal
pattern angular shaped with coarse striae and high dorsal arch, lateral field obscure.

Male

Form I. Labial disc elevated, lateral lips present, head relatively high and slightly set off from
body. Long stylet, knobs backwardly sloping and ovoid shaped. DGO to stylet knobs relatively
long. Lateral field areolated, fifth incomplete incisure present.
Form II. Head clearly set off from the body, lateral lips not present. Head cap rounded. Stylet
delicate and medium-sized, knobs small rounded and set off from shaft. Lateral field with four
incisures, outer bands areolated.

Second-stage juvenile

Form I. Body relatively stout. Hemizonid posterior to S-E pore. Tail conical shaped with
relatively short tail and hyaline tail part. Rectum not inflated. Anterior region of hyaline tail part
with ventral indent in epidermis.
Form II. Body relatively smaller and shorter. DGO to stylet knobs relatively long. Hemizonid
anterior to S-E pore. Tail medium sized, narrowly tapering, terminus finely rounded, sometimes
club shaped. Hyaline tail part not clearly delimitated in anterior region. Rectum inflated.

Hosts

Detected on *Rosa* sp., *Solanum nigrum* L. and *Fraxinus excelsior* L. at the type locality. *Daucus
carota* L., *Arachis hypogaea* L., *Capsicum frutescens* L. and *Lycopersicon esculentum* Mill. were
found to be hosts in an additional greenhouse test. *M. deconincki* induced on all these plants
small galls (Elmiligy, 1968).

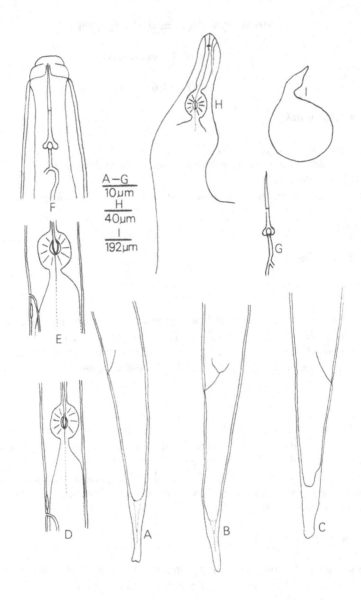

Fig. 4. *Meloidogyne deconincki* (lateral). A-E: Second-stage juvenile. A,B: Tail form II; C: Tail form I; D: Metacorpus region form II; E: Metacorpus region form I; F: Male head end form I (lateral); G-I: Female (lateral). G: Stylet; H: Anterior end; I:Body shape.

Distribution

Only reported from the type locality.

Type locality & host

Described from the tree *Fraxinus excelsior* L., from the garden of an old university building in the centre of Gent Belgium.

Type material

All slides (holo- and paratypes) mentioned in the original description and deposited at Gent University Belgium, were studied.

Etymology

This species was named after the late nematologist Prof. dr. L. De Coninck.

Remarks

The female holotype and the form I male and juvenile paratypes, are in agreement with *M. ardenensis*, while the form II male and juveniles strongly resemble *M. hapla*.

Jepson (1987) reported about the type locality: 'Type locality of *M. deconincki* no longer excist'. This is not correct and needs some explanation. Soon after the species description the type locality was restored, most of the plants and soil were removed (A. Goemaes, pers. comm.). The old *Fraxinus excelsior* in the garden centre however, was untouched and is still present 30 years after the description. Soil and root samples, taken in 1994 and 1995 on and around the type host, contained only specimens similar to *M. ardenensis*. *M. hapla* was not found; possibly it was removed during the restoration of the garden.

The samples were also used for isozyme tests, exactly the same esterase and malate dehydrogenase isozyme patterns were detected as for *M. ardenensis* (Fig. 6).

In an additional host range test with *M. ardenensis*, *M. deconincki* and *M. hapla* on *Daucus carota*, *Fraxinus excelsior* seedlings, *Lycopersicon esculentum* and *Triticum eastivum* L.. only *M. ardenensis* and *M. deconincki* reproduced on *F. excelsior*, while *M. hapla* reproduced on *D. carota* and *L. esculentum*.

M. deconincki is not different in morphology, host reaction and tested isozymes from *M. ardenensis*. It is reasonable to conclude that the description of *M. deconincki* was based on a *M. ardenensis/ hapla* mixture. Jepson (1987) placed already *M. ardenensis* and *M. deconincki* in one group of species ' with overlapping hosts and very similar morphology'. Both species were described in 1968, but *M. ardenensis* one Nematologica issue before *M. deconincki*. The latter is therefore considered as a junior synonym of *M. ardenensis*.

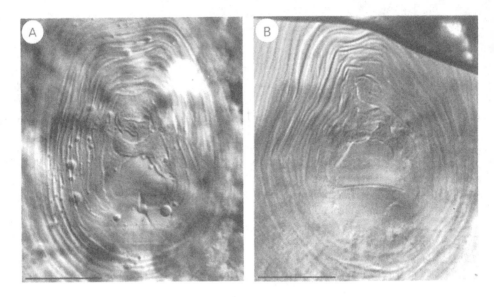

Fig. 5. *Meloidogyne deconincki*. A & B: Female perineal patterns. Bar= 25 µm.

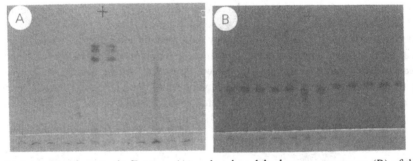

Fig. 6. *Meloidogyne deconincki*. Esterase (A) and malate dehydrogenase patterns (B) of the type populations from Belgium (Gent); reference: *M. javanica* (China) in the two middle positions of the gel.

Table II

Morphometrics of Meloidogyne deconincki *Elmiligy, 1968*
[mean ± SD (range); all measurements in μm]

Character	J2		Males		Females
Form	I	II	I	II	holotype
N	1	5	3	1	
Body length	416	389±17.6 (368-410)	1877±158 (1696-1984)	1174	707
Greatest body diameter	19.6	15.3±0.8 (14.5-15.8)	31.0±1.1 (29.7-31.6)	22.1	422
Body diam. at stylet knobs	-	-	17.3±1.6 (15.8-19.0)	12.6	-
„ „ „ S-E pore	15.2	14.3±1.0 (13.3-15.8)	26.1±1.4 (25.3-27.8)	19.0	-
„ „ „ anus	10.1	11.1±0.7 (10.1-10.7)	-	-	-
Stylet length	12.6	10.2±0.3 (10.1-10.7)	24.1±0.7 (23.4-24.7)	20.2	17.1
DGO	3.8	3.4±0.3 (3.2-3.8)	6.1±0.4 (5.7-6.3)	4.4	6.3
S-E pore to anterior end	-	-	-	-	15.2
Anterior end to metacorpus	-	-	90±4.8 (85-95)	78	81
Metacorpus length	-	-	-	-	34.1
„ diameter	-	-	-	-	31.0
Tail length	39.8	52.1±2.0 (48.7-53.7)	-	-	-
Tail terminus length	12.6	11.9±1.4 (10.7-13.9)	-	-	-
Anus-primordium	107	90±6.1 (85-99)	-	-	-
Spicule	-	-	38.0±4.5 (34.8-41.1)	28.4	-
Gubernaculum	-	-	9.5±0.4 (9.0-10.1)	8.2	-
a	21.2	25.4±0.6 (24.7-26.3)	64±1.8 (62-66)	53	-
b"	6.6	7.6±0.2 (7.3-7.8)	-	-	-
c	10.5	7.5±0.5 (7.0-8.5)	-	-	-
c'	3.9	4.7±0.4 (4.1-5.2)	-	-	-
T	-	-	52±6.5 (47-59)	79	-
(S-E pore/L)x 100	17.5	20.3±0.4 (19.8-20.8)	9.1±0.5 (8.6-9.5)	10.1	

Meloidogyne exigua (Göldi, 1892) Chitwood, 1949

Meloidogyne exigua

Göldi (1892); *Arch. Mus. Nac., Rio de Janeiro* 8: 7-112.
  Chitwood (1949); *Proc. Helminth. Soc. Washington* 16: 90-104.
  Lordello & Zamith (1958); *Proc. Helminth. Soc. Washington* 25: 133-137.
  Whitehead (1968); *Transactions of the Zoological Society of London* 31: 263-401.
  Cain (1974); *Meloidogyne exigua. C.I.H. Descriptions.* Set 4. No. 49. CAB, St. Albans, UK.
  Jepson (1987); *Identification of root-knot nematodes* (Meloidogyne *species*). Wallingford, UK:
  CAB International. 265 pp.
  Eisenback & Hirschmann (1991); In: *Manual of agricultural nematology*, Ed. W.R. Nickle.
  New York: Marcel Dekker. pp: 191-274.

Measurements & Description

The morphology and morphometrics of *M. exigua*, the type species of the genus, was described
in detail by Cain (1974) and more recently by Eisenback and Hirschmann (1991).

Hosts

The main host is *Coffea arabica* L., also other, less important, dicotyledonous hosts are known
(Jepson, 1987).

Distribution

South America and the Caribbean, incidentily reported from Thailand, India, Greece and Italy.

Type locality & host

Province of Rio de Janeiro Brazil, described from *Coffea* sp.

Type material

The Chitwood (1949) slides were not available for us. Only one slide with developing males and
females from Peru, used by Whitehead (1968), was available.

Etymology

The species epithet means 'little' or 'small'.

Remarks

  M. exigua has been reported three times from Europe, first on *Lycopersicon esculentum* Mill.
from Italy (Sconamiglio, 1968), then on *Prunus persica* Stokes from Greece (Kaliopanos, 1978),
and finally on *Bougainvillea glabra* Mill. from Italy (Scognamiglio *et al.*, 1985). These are all
reports of local plant infections; there is no indication that *M. exigua* is wide spread in southern

Europe. It is more likely that *M. exigua* was introduced, for instance the ornamental *Bougainvillea* originates from Brazil, where *M. exigua* is widely distributed.

Esbenshade & Triantaphyllou (1985) described the esterase (VF1 type) and malate dehydrogenase (N1 type) patterns for *M. exigua*.

Meloidogyne hapla Chitwood, 1949

(Fig. 7 & 8)

Meloidogyne hapla

Chitwood (1949); *Proc. Helminth. Soc. Washington* 16: 90-104.
  Gillard (1961); *Meded. LandbHoogesch., Gent* 26: 515-646.
  Whitehead (1968); *Transactions of the Zoological Society of London* 31: 263-401.
  Orton Williams (1974); *Meloidogyne hapla. C.I.H. Descriptions.* Set 3, No.31. CAB, St. Albans, UK.
  Sturhan (1976); *Nachrichtenbl. Deut. Pflanzenschutzd.* 28: 113-117.
  Eisenback & Hirschmann (1979); *Journal of Nematology* 11: 5-16.
  Jepson (1987); *Identification of root-knot nematodes* (Meloidogyne *species*). Wallingford, UK: CAB International. 265 pp.
  Eisenback & Hirschmann (1991); In: *Manual of agricultural nematology*, Ed. W.R. Nickle. New York, Marcel Dekker. pp: 191-274.
  Eisenback (1993); *Fundam. Appl. Nematol.* 16: 259-271.

Measurements & Description

The above mentioned references are a selection of the most important taxonomical literature. Orton Williams (1974) and Eisenback and Hirschmann (1991) are very useful references for a complete treatise on *M. hapla* morphology and morphometrics.

Hosts

Goodey *et al.* (1965) published an impressive list of dicotyledonous plants parasitized by *M. hapla*, including economic important food crops and ornamentals. It has only been reported two times from monocotyledons; *Allium* sp. (Oostenbrink, 1960) and *Hosta* sp. (Brinkman & Goossens, 1994). It induces small galls, often with secondary roots.

Distribution

In Europe *M. hapla* is undoubtedly the most common and widely distributed root-knot nematode in agricultural areas; it is also detected in natural habitats like coastal dunes, woods and river banks.

Type locality & host

Bridgehamton, Long Island, New York USA on *Solanum tuberosum* L.

Type material

One lectotype with female heads and perineal patterns from the type locality (Whitehead, 1968) was available for study.

Etymology

The name *hapla* probably refers to the obsolete word 'haply', meaning 'by change' or 'maybe'.

Remarks

Two cytological forms are described for *M. hapla*, race A is ranging in chromosome number from 14 to 17 and reproduces by facultative meiotic parthenogenesis, while race B is polyploid and reproduces by mitotic parthenogenesis (Triantaphyllou, 1985). Both races differ in second-stage juvenile tail and body length but are negligibly different in other morphological characters (Eisenback and Hirschmann, 1991; Eisenback, 1993). Both races occur in the Netherlands, with race A most prevalent (Brinkman and Goosens, 1994), but little is known about the distribution of both forms elsewhere in Europe.

All the studied populations, including race A & B, show a H1 malate dehydrogenase and a H1 esterase type.

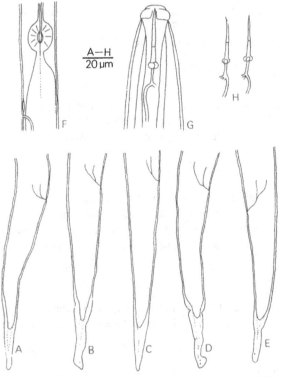

Fig. 7. *Meloidogyne hapla*. A-F: Second-stage juvenile (lateral). A-E: Tails; F: Metacorpus region; G: Male anterior end (lateral); H: Female stylets (lateral).

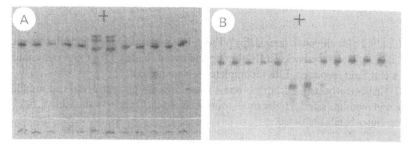

Fig. 8. *Meloidogyne hapla.* Esterase (A) and malate dehydrogenase patterns (B) of populations from the Netherlands (Bavel and Ammerzoden), Germany (Köln), France (Concarneu) and Hungary, populations from Bulgaria, Norway, Poland and UK with the same types were not included; reference: *M. javanica* (China) in the two middle positions of the gel.

Meloidogyne litoralis Elmiligy, 1968

[= Syn. with *M. ardenensis*]

(Fig. 9 & 10)

Meloidogyne litoralis

Elmiligy (1968); *Nematologica* 14:577-590.
 Jepson (1987); *Identification of root-knot nematodes* (Meloidogyne *species*). Wallingford, UK: CAB International. 265 pp.

Measurements

(in glycerine) in Table III

Female

Holotype. Body pear shaped, relatively small with long curved neck and thickening behind metacorpus. Long stylet, cone slightly curved dorsally, knobs ovoid and backwardly sloping. S-E pore less than one stylet length from head end. Distance DGO to stylet knobs relatively long. Perineal pattern angular shaped with coarse striae and high dorsal arch, lateral field obscure. *Deviating perineal patterns in some paratypes.* Patterns rounded with fine striae and low dorsal arch, lateral field visible, punctations present in tail terminus area, in some patterns ventral striae extend laterally (Fig. 10B).

Male

Form I. Labial disc elevated, lateral lips present, head relatively high and slightly set off from body. Long stylet, knobs backwardly sloping and ovoid shaped. DGO to stylet knobs relatively long. Lateral field areolated, fifth incomplete incisure present.
Form II. Head clearly set off from the body, lateral lips not present. Head cap rounded. Stylet delicate and medium-sized, knobs small rounded and set off from shaft. Lateral field with four incisures, outer bands areolated.

Second-stage juvenile

Body moderately long and relatively stout. Hemizonid posterior to the S-E pore. Pharyngeal gland lobe well developed, ventrally overlapping of intestine sometimes difficult to observe. Tail conical with relatively short tail and hyaline tail part. Rectum rarely inflated. Anterior region of hyaline tail part with ventral indent in epidermis. Hyaline tail part with cuticular constriction, ending in rounded terminus.

Hosts

No other hosts are known beside the type host *Ligustrum* sp., where it induced small galls.

Distribution

Only known from the type locality: dunes near Ambleteuse, Pas de Calais France.

Type material

All the slides (holo- and paratypes) as mentioned in the description, deposited at Gent University Belgium, were studied.

Etymology

The species names means 'beach' or 'shore inhabitant'.

Remarks

The female holotype and the form I male and juvenile paratypes are in agreement with *M. ardenensis*, while the form II males and some female perineal patterns fit with *M. hapla*. We were not able to study living type material, as the precise place of sampling at the type location area is unknown. The description does not mention the second-stage juvenile hemizonid position for *M. litoralis*. A posterior position to the S-E pore was observed, i.e. the same hemizonid position as described for *M. ardenensis* second-stage juveniles.

It is not unusual to find *M. hapla* or *M. ardenensis* on *Ligustrum vulgare* L., as both were reported from this host plant (Santos, 1968; Coolen *et al.*, 1975). Both species were also detected together on *Ligustrum vulgare* in the coastal dunes near Oostvoorne, the Netherlands (Karssen, unpub. results). As with *M. deconincki* and *M. ardenensis*, Jepson (1987) placed also *M. litoralis* in a group of related species parasitizing Oleaceae.

I consider *M. litoralis* as a junior synonym of *M. ardenensis*.

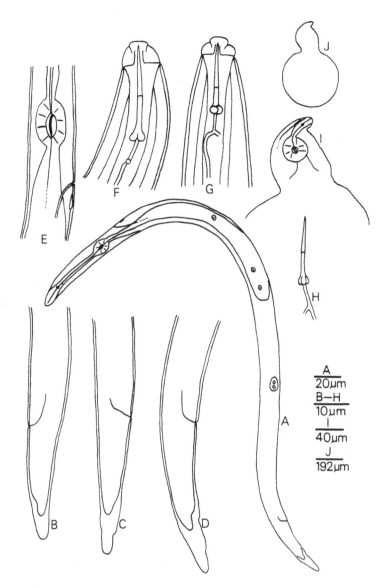

Fig. 9. *Meloidogyne litoralis*. A-E: Second-stage juvenile (lateral). A: Body; B-D: Tails; E: Metacorpus region; F,G: Male. F: Head end form I (dorsal); G: Head end form II (ventral); H-J: Female (lateral). H: Stylet; I: Anterior end; J: Body shape.

Table III

Morphometrics of Meloidogyne litoralis *Elmiligy, 1968*
[mean ± SD (range); all measurements in μm]

Character	J2	Males I	Males II	Females holotype
Form		I	II	holotype
N	10	1	1	
Body length	402±8.0 (390-416)	1216	1146	614
Greatest body diameter	15.2±1.7 (13.9-18.3)	41.7	33.5	400
Body diam. at stylet knobs	-	18.3	17.7	-
,, ,, ,, S-E pore	13.2±1.0 (12.6-15.8)	25.9	26.5	-
,, ,, ,, anus	9.9±0.3 (9.5-10.1)	-	-	-
Stylet length	11.4±0.6 (10.1-12.0)	24.0	20.2	17.7
DGO	3.5±0.3 (3.2-3.8)	5.7	4.4	5.7
S-E pore to anterior end	-	-	-	13.3
Anterior end to metacorpus	-	95	78	63
Metacorpus length	-	-	-	34.8
,, diameter	-	-	-	30.3
Tail length	39.2±2.4 (34.8-44.2)	-	-	-
Tail terminus length	11.2±1.4 (7.6-12.0)	-	-	-
Anus-primordium	101±4.3 (95-107)	-	-	-
Spicule	-	34.1	32.9	-
Gubernaculum	-	9.5	8.2	-
a	26.8±2.7 (21.3-29.0)	29.2	34.2	-
b"	6.7±0.3 (6.4-7.3)	-	-	-
c	10.3±0.7 (8.8-11.4)	-	-	-
c'	4.0±0.3 (3.7-4.7)	-	-	-
T	-	50	52	-
(S-E pore/L)x 100	19.6±1.5 (17.8-21.8)	10.6	10.6	-

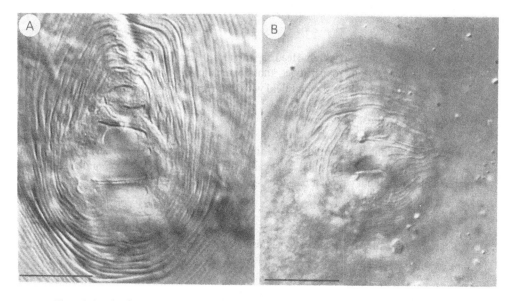

Fig. 10. *Meloidogyne litoralis*. A & B: Female perineal patterns. Bar= 25 μm.

Meloidogyne lusitanica Abrantes & Santos, 1991

(Fig. 11 & 12)

Meloidogyne lusitanica

Abrantes & Santos (1991); *Journal of Nematology* 23: 210-224.
  Abrantes & Vovlas (1988); *Canadian Journal of Zoology* 66: 2852-2854.
  Pais & Abrantes (1989); *Journal of Nematology* 21: 342-346.
  Abrantes *et al.* (1992); *Nematologica* 38: 320-327.

Measurements

(in glycerine) see Table IV

Female

Body apple to pear shaped, with short neck, posteriorly rounded. Relatively long stylet, cone slightly curved dorsally, knobs ovoid and backwardly sloping. S-E pore between stylet knobs and metacorpus level. Striking perineal pattern, trapezoid shaped with coarse striae and medium to high dorsal arch, lateral field weakly visible.

Male

Head region set off, relatively high. Head cap rounded, lateral lips absent. Head region with incomplete incisures. Long robust stylet, knobs ovoid shaped and backwardly sloping. DGO to stylet knobs relatively long. Lateral field with four incisures, outer bands areolated.

Second-stage juvenile

Body moderately long. Hemizonid ranging from level of excretory pore to anteriorly adjacent. Tail conical with relatively short tail and hyaline tail part. Rectum not always inflated. Hyaline tail part without cuticular constriction, ending in rounded terminus.

Hosts

No other hosts are known, beside the type host *Olea europaea* L.

Distribution

So far only detected in Portuguese olive fields, where it was found in 2 of the 90 surveyed fields (Abrantes & Santos, 1991).

Type locality

A field near Cadaixo, Miranda do Corvo Portugal.

Type material

The paratypes deposited at Rothamsted Experimental Station, Harpenden UK, were studied.

Etymology

The name refers to the Roman province that included most of present Portugal.

Remarks

The studied slides are in agreement with the description, except that the described punctations in the perineal patteren tail terminus area were not present in the observed patterns.

Pais and Abrantes (1989) described the esterase (P1 type) and malate dehydrogenase (P3 type) patterns for *M. lusitanica*. If one correlates these types to the original isozyme coding types of Esbenshade and Triantaphyllou (1985) and exclude for Mdh the minor bands, one could rename these types as A1 for esterase and N1c, a new type, for malate dehydrogenase.

M. lusitanica is morphologically related to *M. ardenensis*. Both also share one host group, the family Oleaceae, which includes for example *Olea europaea*, *Ligustrum vulgare* L. and *Fraxinus excelsior* L.

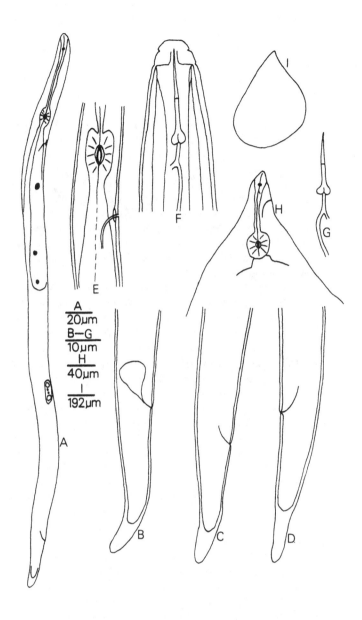

Fig. 11. *Meloidogyne lusitanica.* A-E: Second-stage juvenile (lateral). A: Body; B-D: Tails; E: Metacorpus region; F: Male head end (lateral); G-I: Female (lateral). G: Stylet; H: Anterior end; I: Body shape.

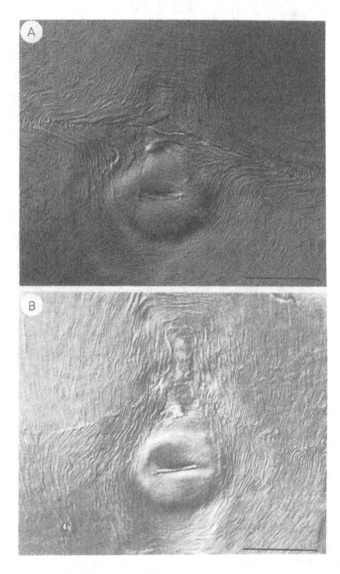

Fig. 12. *Meloidogyne lusitanica*. A & B: Female perineal patterns. Bar= 25 µm.

Table IV

Morphometrics of Meloidogyne lusitanica *Abrantes & Santos, 1991*
[mean ± SD (range); all measurements in μm]

Character	J2	Males	Females
N	5	1	1
Body length	394±10.0	1728	499
	(384-406)		
Greatest body diameter	15.2±0.5	37.9	666
	(14.5-15.8)		
Body diam. at stylet knobs	-	19.6	-
,, ,, ,, S-E pore	13.7±0.3	28.4	-
	(13.3-13.9)		
,, ,, ,, anus	11.1±0.4	-	-
	(10.7-11.4)		
Stylet length	13.8±0.3	24.6	17.6
	(13.3-13.9)		
DGO	3.7±0.3	5.1	4.8
	(3.2-3.8)		
S-E pore to anterior end	-	-	40.0
Anterior end to metacorpus	-	104	98
Metacorpus length	-	-	44.8
,, diameter	-	-	41.6
Tail length	40.4±4.0	-	-
	(35.0-45.5)		
Tail terminus length	11.7±1.7	-	-
	(10.1-14.5)		
Anus-primordium	93±2.7	-	-
	(89-95)		
Spicule	-	37.9	-
Gubernaculum	-	10.1	-
a	26.0±0.7	45.6	-
	(25.3-26.7)		
b''	6.7±0.1	-	-
	(6.6-6.8)		
c	9.8±0.8	-	-
	(8.9-11.1)		
c'	3.6±0.3	-	-
	(3.3-4.0)		
T	-	51	-
(S-E pore/L)x 100	20.2±0.6	10.0	-
	(19.3-20.9)		

Meloidogyne megriensis (Pogosjan, 1971) Esser, Perry & Taylor, 1971

Syn. *Hypsoperine megriensis* Pogosjan, 1971
   Hypsoperine (*Hypsoperine*) *megriensis* (Pogosjan, 1971) Siddiqi, 1986

Meloidogyne megriensis

Pogosjan (1971); *Doklady Akademmii Nauk Armyanskoi SSR* 53: 306-312.
   Esser *et al.* (1976); *Proc. Helminth. Soc. Washington* 43: 138-150.
   Jepson (1987); *Identification of root-knot nematodes* (Meloidogyne *species*). Wallingford,
   UK: CAB International. 265 pp.
   Siddiqi (1986); *Tylenchida. Parasites of Plants and Insects.* St. Albans, UK: CAB. 645 pp.

Measurements & Description

The only description available is the Russian description of Pogosjan (1971), Jepson (1987)
mentioned some of the most important (translated) diagnostic characters.

Hosts

No other hosts are known beside the type host *Mentha longifolia* (L.) Huds.

Distribution

Only known from the type locality, an orchard in Megri, and a nearby village named
Vagravar, Armenia, on 700-1000 masl.

Type material

Type material, deposited in Armenia, or living material was not available, due to the recent
war activities.

Etymology

The species name refers to Megri the name of the type locality.

Remarks

   The original description of *M. megriensis* (a Russian-Dutch translation), is rather
incomplete. Together with the non-availability of types and/or living material it is very
difficult to clarify the status of this species. There is some morphological resemblance to *M.
hapla.*
   Eisenback and Hirschmann (1991) also had doubts about the species status of *M.
megriensis*, and placed it under *nomen nudum*. However, as types were deposited and a
description was formally published, it is better to name it *species inquirenda* for the moment.

Discussion

This revision reduces the number of *Meloidogyne* species parasitizing on dicotyledonous hosts in Europe mainly to three species, i.e. *M. ardenensis*, *M. hapla* and *M. lusitanica*. New species descriptions will increase this number in the future. Particularly in natural areas one may expect to find more new forms. For example, Dr. D. Sturhan discovered undescribed *Meloidogyne* second-stage juveniles near the roots of *Salix* sp. on several locations near the river Rhine in Germany.

Together we sampled in the river bank of the Rhine near Köln and discovered a mixture of *M. ardenensis*, *M. hapla* and the undescribed *Meloidogyne* species on *Salix* sp. The complete description of this new form is retarded, due to this complex and the fact that we are unable to culture it so far. The second-stage juveniles and females of this new form are morphologically related with *M. caraganae* Shagalina *et al.*, 1985 and *M. turkestanica* Shagalina *et al.*, 1985. I have found this complex of *Meloidogyne* species also in the Netherlands, near Kampen at the IJssel river, a sidebranch of the river Rhine. Possibly the river Rhine plays an important role in the distribution of these plant parasites. Coincidentally this new form has the same malate dehydrogenase pattern as described for *M. fallax*, but differs in esterase pattern.

With the synonymization of *M. deconincki* and *M. litoralis* with *M. ardenensis*, Dr. I.A. Elmiligy actually detected *M. ardenensis* for the first time on continental Europe. The observation that both junior synonyms are based on mixtures of *M. hapla* and *M. ardenensis* indirectly proofs that species mixtures of these species are not only restricted to the river Rhine, but likely to be more widespread. Possibly other *Meloidogyne* species, either not tested for purity before description or not based on a single egg-mass culture, will be unravelled as a species mixture.

Although the studied species mainly parasitize one host group, they are morphologically distinct but not so closely related as within the 'graminis-group' (Jepson, 1987; chapter 4). *M. ardenensis* and *M. lusitanica* share a few characters as a robust, long male and female stylet; a relatively short juvenile tail and hyaline tail length, and well developed pharyngeal glands. *M. ardenensis*, *M. lusitanica* and *M. hapla* have one common character in all stages, a relatively posterior DGO. The studied group of European species parasitizing on dicotyledons, however is too small to recognize morphological trends or to define a 'dicotyledon-group' based on morphological characteristics.

A robust and long male stylet is, for a non-feeding stage with a reduced anterior pharyngeal gland nucleus, rather curious. Hypothetically the male stylet could be a useful or even a necessary tool to leave the root after the last molt.

M. hapla juvenile tails are variable in shape, particularly in the posterior part (Fig. 7A-E). This hyaline tail part is ranging from finely rounded to pointed, with aberrant tips (subacute to bifid) observed in most populations. In combination with an indistinct anterior hyaline tail part, it partly explains the wide range in tail and hyaline tail length observed by several authors.

The moderate quality of the studied type material of *M. ardenensis*, *M. deconincki* and *M. litoralis* was an additional obstacle in the attempt to clearify the taxonomical position of these species. Although these types were useful, after careful and prolonged observation, it is not unlikely that this material will be useless within a few years. Poor quality type slides were also noticed for other European *Meloidogyne* species.

References

ABRANTES, I.M. de O. & SANTOS, M.S.N. de A. (1991). *Meloidogyne lusitanica* n. sp. (Nematoda: Meloidogynidae), a root-knot nematode parasitizing olive tree (*Olea europaea* L.). *J. Nematol.* 23, 210-224.

BRINKMAN, H. & GOOSSENS, J. (1994). Het noordelijk wortelknobbelaaltje *Meloidogyne hapla* bij Hosta-soorten. *Gewasbescherming* 25, 79-82.

BRZESKI, M.W. (1996). [Root-knot nematodes in field crops]. *Ochrona Roslin* 40, 4-5.

CAIN, S.C. (1974). *Meloidogyne exigua. C.I.H. Descriptions of plant-parasitic nematodes.* Set 4, No. 49, St. Albans, UK, C.A.B.

CHITWOOD, B.G. (1949). 'Root-knot nematodes'. Part 1. A revision of the genus *Meloidogyne* Goeldi 1887. *Proc. Helminth. Soc. Wash.* 16, 90-104.

CHIZOV, V.N., TURKINA, A. YU. (1986). [*Meloidogyne ardenensis* a parasite of *Betula pendula* in Moscow region]. *Byulleten Vsesoyuznogo Instituta Gelmintologii im. K.I. Skryabina* 45, 100-102.

COOLEN, W.A., HENDRICKX, G. & D'HERDE, C.J. (1975). *Waardplantonderzoek in de boomteelt, als basis van een mogelijke vruchtafwisseling ter bestrijding van nematoden.* Publikatie nr. W 17, Rijksstation voor Nematologie en Entomologie, Merelbeke, Belgium. 33 pp.

EISENBACK, J.D. & HIRSCHMANN, H. (1991). Root-knot nematodes: *Meloidogyne* species and races. In: *Manual of agricultural nematology.* pp. 191-274. Ed. W.R. Nickle. New York, Marcel Dekker.

EISENBACK, J.D. (1993). Morphological comparisons of females, males and second-stage juveniles of cytological races A and B of *Meloidogyne hapla* Chitwood, 1949. *Fundam. Appl. Nematol.* 16, 259-271.

ELMILIGY, I.A. (1968). Three new species of the genus *Meloidogyne* Goeldi, 1887 (Nematoda: Heteroderidae). *Nematologica* 14, 577-590.

ESBENSHADE, P.R. & TRIANTAPHYLLOU, A.C. (1985). Use of enzyme phenotypes for identification of *Meloidogyne* species. *J. Nematol.* 17, 6-20.

ESBENSHADE, P.R. & TRIANTAPHYLLOU, A.C. (1987). Enzymatic relationships and evolution in the genus *Meloidogyne* (Nematoda: Tylenchida). *J. Nematol.* 19, 8-18.

ESSER, R.P., PERRY, V.G. & TAYLOR, A.L. (1976). A diagnostic compendium of the genus *Meloidogyne* (Nematoda: Heteroderidae). *Proc. Helminth. Soc. Wash.* 43, 138-150.

FRANKLIN, M.T. (1957). Review of the genus *Meloidogyne. Nematologica* 2 (Suppl.), 387-397.

FRANKLIN, M.T. (1972). The present position in the systematics of *Meloidogyne*. *OEPP/EPPO Bull.* 6, 5-15.

FRANKLIN, M.T. (1979). Taxonomy of the genus *Meloidogyne*. In: *Root-knot nematodes* (Meloidogyne *species*). pp. 37-54. Eds. F. Lamberti & C.E. Taylor. London, New York, San Francisco, Academic Press.

GOFFART, H. (1957). Bemerkungen zu einigen Arten der Gattung *Meloidogyne*. *Nematologica* 2, 177-184.

GÖLDI, E.A. (1892). Relatoria sôbre a molestia do cafeiro na provincia da Rio de Janeiro. *Archos Mus. nac., Rio de Janeiro* 8, 7-112.

GOODEY, J.B., FRANKLIN, M.T. & HOOPER, D.J. (1965). *The nematode parasites of plants catalogued under their hosts*. 3rd edn. Farnham House, Farnham Royal, England, C.A.B. 214 pp.

HEWLETT, T.E. & TARJAN, A.C. (1983). Synopsis of the genus *Meloidogyne* Goeldi, 1887. *Nematropica* 13, 79-102.

HIRSCHMANN, H. (1985). The genus *Meloidogyne* and morphological characters differentiating its species. In: *An advanced treatise on* Meloidogyne, *Volume I*. pp. 79-94. Eds. J.N. Sasser & C.C. Carter. Raleigh, NC, USA, North Carolina State University Graphics.

JEPSON, S.B. (1987). *Identification of root-knot nematodes* (Meloidogyne *species*). Wallingford, UK, C.A.B. International. 265 pp.

KALIOPANOS, C.N. (1978). The problem of the root-knot nematodes in Greece. In: *Proceedings of the research planning conference on root-knot nematodes*, Meloidogyne *spp.* pp. 20-24. Cairo, Giza, Egypt.

KARSSEN, G., VAN HOENSELAAR, T., VERKERK-BAKKER, B. & JANSSEN, R. (1995). Species identification of root-knot nematodes from potato by electrophoresis of individual females. *Electrophoresis* 16, 105-109.

KARSSEN, G. (1996). Description of *Meloidogyne fallax* n. sp. (Nematoda: Heteroderidae), a root-knot nematode from The Netherlands. *Fundam. Appl. Nematol.* 19, 593-599.

KARSSEN, G., VAN AELST, A.C. & VAN DER PUTTEN, W.H. (1998). *Meloidogyne duytsi* n. sp. (Nematoda: Heteroderidae), a root-knot nematode from Dutch coastal foredunes. *Fundam. Appl. Nematol.* 21, 299-306.

KLEYNHANS, K.P.N. (1989). *The root-knot nematodes of South-Africa*. Pretoria, South Africa, Plant Protection Research Institute. 61 pp.

OOSTENBRINK, M. (1960). Enige bijzondere aaltjesaantastingen in 1959. *T. Pl. Ziekten* 67, 126-127.

ORTON WILLIAMS, K.J. (1974). *Meloidogyne hapla. C.I.H. Descriptions of plant-parasitic nematodes.* Set 3, No. 31, St. Albans, UK, C.A.B.

PAIS, C.S. & ABRANTES, I.M. de O. (1989). Esterase and malate dehydrogenase phenotypes in Portuguese populations of *Meloidogyne* species. *J. Nematol.* 21, 342-346.

POGOSJAN, E.E. (1971). *Hypsoperine megriensis* n. sp. (Nematoda: Heteroderidae) in the Armenian SSR. *Doklady Akademii Nauk Armyanskoi SSR* 53, 306-312.

RICHTER, M. (1981). Ein erster Nachweis von *Meloidogyne ardenensis* Santos, 1968, in der DDR. *Nachrichtenbl. Pflanzenschutzd. DDR* 35, 33-36.

SANTOS, M.S.N. de A. (1968). *Meloidogyne ardenensis* n. sp. (Nematoda: Heteroderidae), a new British species of root-knot nematode. *Nematologica* (1967) 13, 593-598.

SCOGNAMIGLIO, A. (1968). Prove di lotta contro nematodi galligeni delle radici su pomodoro, con l'impiego di formulati granulari sistemici. *Boll. Ent. Agr. Filippo Silvestri, Portici (Napoli).* 26, 1-20.

SCOGNAMIGLIO, A., BIANCO, M. & MARULLO, R. (1985). *Meloidogyne exigua* Goeldi su radici di *Bougainvillea glabra* 'Sanderiana'. *L'Informatore Agrario, Verona* 16, 63-65.

THOMAS, P.R. & BROWN, D.J.F. (1981). Some hosts of a *Meloidogyne ardenensis* population found in Scotland. *Pl. Pathol.* 30, 147-151.

TRIANTHAPHYLLOU, A.C. (1985). Cytogenetics, cytotaxonomy and phylogeny of root-knot nematodes. In: *An advanced treatise on* Meloidogyne, *Volume I.* pp. 113-126. Eds. J.N. Sasser & C.C. Carter. Raleigh, NC, USA, North Carolina State University Graphics.

WHITEHEAD, A.G. (1968). Taxonomy of *Meloidogyne* (Nematoda: Heteroderidae) with descriptions of four new species. *Trans. Zool. Soc. Lond.* 31, 263-401.

4

REVISION OF THE

EUROPEAN ROOT-KNOT NEMATODES

II. ON MONOCOTYLEDONS

'De natuur bevat op zichzelf geen logica, maar als wij de natuur beschrijven willen, moeten wij dit logisch doen, op straffe van het helemaal niet te doen.'

W. F. Hermans (in 'Wittgenstein', 1990).

Introduction

This second part of the revision of the European root-knot nematodes (*Meloidogyne* Göldi, 1892) is focused on *Meloidogyne* species parasitizing on Poaceae and Cyperaceae field hosts.

According to Jepson (1987), it is the most well defined group within the genus. She named this cluster of species the 'graminis-group', after *M. graminis* (Sledge & Golden, 1964) the first described species in this group. The 'graminis-group' shares the following biological characteristics, as defined by Jepson (1987):

1) In the field they are confined to Poaceae and/or Cyperaceae hosts.
2) The female body is markedly elongate.
3) In the female the vulva is situated on a posterior protuberance (exception, *M. sewelli*).
4) Often confined to, or more often found in damp or wet conditions (exceptions, *M. graminis* and *M. marylandi*).
5) Galling of the host roots is slight or absent.
6) Key morphological characters for the group have dimensions within narrow limits.
7) Males common in field soil.
8) Where the mode of reproduction is known it is facultatively parthenogenetic.

World-wide this group includes about twenty species, in Europe the following species have been detected so far: *M. duytsi, M. graminis, M. kralli, M. maritima* and *M. naasi. M. duytsi* and *M. maritima* were recently described respectively redescribed (Karssen *et al*, 1998a & b).

Material and Methods

Types and field populations of *M. duytsi, M. graminis, M. kralli, M. maritima* and *M. naasi* were studied, as described in chapter 3. The following procedure was applied for description of *M. duytsi* and redescription of *M. maritima*.

Before morphological and morphometric studies, the nematode population was first tested for purity with isozyme electrophoresis based on esterase and malate dehydrogenase of young egg-laying females (Esbenshade & Triantaphyllou, 1985; Karssen *et al.*, 1995). Adult females were hand-picked from infected roots, while males and second-stage juveniles (J2) were extracted by rinsing fresh root samples in a spray mist chamber for one week.

For light microscope (LM) studies eggs, J2's, males and females (head regions and perineal patterns) were fixed at 70° C, and mounted in TAF (Courtney *et al.*, 1955). Measurements and photographs were taken with a light microscope (Olympus BH2) using differential interference contrast (DIC).

For scanning electron microscope (SEM) observations males and females were fixed in 3% glutaraldehyde buffered with 0.05 M phosphate buffer (pH 6.8) for 1.5 h and post-fixed with 2% osmium tetroxide for 2 h at 22 °C. The specimens were dehydrated in a seven-graded series of ethanol (Wergin, 1981), critical-point dried with carbon dioxide, and sputter coated with 4-5 nm Pt in a dedicated preparation chamber (CT 1500 HT, Oxford Instruments). The nematodes were examined with a field emission electron microscope (Jeol 6300 F) at 5 kV.

For preparation of type material females, males and J2's were fixed in hot FP [11 ml formalin (40% formaldehyde), 1 ml propionic acid, 88 ml distilled water] and processed by a rapid glycerin-ethanol method (Seinhorst, 1959). The specimens were mounted in desiccated glycerin on Cobb slides.

Species descriptions

Meloidogyne duytsi Karssen, van Aelst & van der Putten, 1998

(Fig. 1-5)

Meloidogyne duytsi

Karssen *et al.* (1998a); *Fundamental and applied Nematology* 21: 299-306.

Preface

In 1976, Sturhan reported the root-knot nematode *M. graminis* (Sledge & Golden, 1964) Whitehead, 1968 for the first time on *Ammophila arenaria* (L.) Link from coastal foredunes in Europe. In 1978 and 1980, Cook found *M. graminis* on various locations in the U.K. on *A. arenaria* (see Jepson, 1987). Later, these populations were described as a new species: *M. maritima* Jepson, 1987. Brinkman (1985) reported the occurrence of *M. graminis* (attributed to *M. maritima* by Maas *et al.*,1987) and an unknown *Meloidogyne* species on *A. arenaria*, sampled by Dr. K. Kuiper in foredunes on the Dutch island of Schiermonnikoog.

A decline of the natural sand-stabilizing plant species *A. arenaria* is extensively studied by ecologists in the foredunes near Oostvoorne, the Netherlands. A disease complex of nematodes and fungi is thought to be responsible for the degeneration of *A. arenaria* (Van der Putten *et al.*, 1993; de Rooij-Van der Goes, 1995, & *et al.*, 1995; Little & Maun, 1996).

In 1996, the senior author received *Elymus farctus* (Viv.) Melderis samples from coastal dunes near Oostvoorne for identification. The samples included an undescribed *Meloidogyne* species very different from *M. maritima* and other root-knot nematodes. This new species was recently described as *Meloidogyne duytsi.*

Measurements

(in TAF) see Table I

Female

Body relatively large, annulated, pearly white, globular shaped, neck region distinct and projecting from the body axis to one side, posterior protuberance not observed. Head region set off from body. Head cap distinct but rather variable in shape, labial disk slightly elevated. Cephalic framework weakly sclerotized.

Stylet cone slightly curved dorsally, shaft cylindrical; large knobs, transversely ovoid and set off from the shaft. Excretory pore located half-way between metacorpus and head end. No vesicles were observed near the lumen lining of the metacorpus. Pharyngeal glands rather variable in shape and size.

Perineal pattern asymmetrical shaped; dorsal arch relatively low, with coarse striae; one lateral wing, variable in size, present in most patterns. Tail terminus distinct and without punctations; lateral lines indistinct. Phasmids small, located just above the covered anus, the inter-phasmidial distance is 20.2 ± 2.7 µm. Ventral pattern region angular shaped with fine striae.

Male

Body vermiform, annulated and twisted. Four incisures present in lateral field, outer ones regularly areolated. Head set off from the body, one relatively high post-labial annule (or head region) present, transverse incisures not observed. Labial disc rounded, slightly elevated and fused with the medial lips into a head cap. Medial lips rounded, lateral edges slightly wider than the labial disc. Prestoma hexagonal in shape with six inner sensilla. Four cephalic sensilla present on the medial lips and marked by small cuticular depressions. Slit-like amphidial openings present between the head cap and the relatively small lateral lips.

Cephalic framework moderately sclerotized, vestibule extension indistinct. Stylet with straight cone and cylindrical shaft; large transversely ovoid knobs, set off from the shaft. Pharynx with slender procorpus and oval shaped metacorpus. Pharyngeal gland lobe ventrally overlapping the intestine, variable in length, two subventral gland nuclei present.

Hemizonid, 4-5 µm long and positioned 5 to 12 µm anterior to the excretory pore. Testis long, monorchic and usually with outstretched germinal zone. Tail twisted, relatively short and pointed. Spicula slender, curved ventrally, two pores present on the spiculum tip. Phasmids located posterior to cloaca.

Second-stage juvenile

Body vermiform, moderately long and annulated. Lateral field with four incisures, weakly areolated. Head region truncate and slightly set off from the body. Cephalic framework weakly sclerotized, vestibule extension indistinct. Stylet moderately long, cone straight, shaft cylindrical; knobs rounded to transversely ovoid.

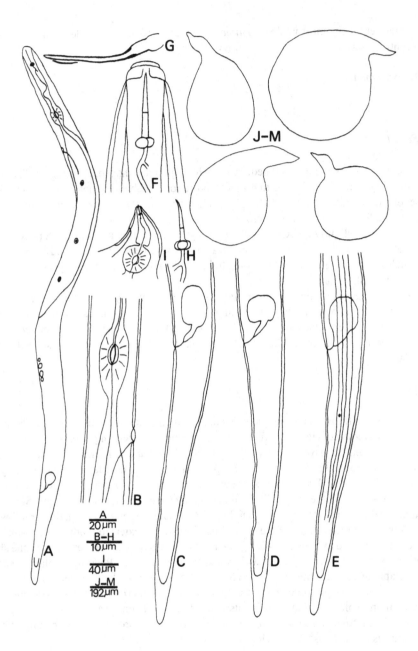

Fig. 1. *Meloidogyne duytsi*. A-E: Second-stage juvenile (lateral). A: Body; B: Metacorpus region; C-E: Tail variation; F-G: Male (lateral). F: Head end; G: Spicule; H-M: Female (lateral). H: Stylet; I: Anterior end ; J-M: Body shapes.

Metacorpus ovoid, triradiate lining moderately sclerotized. Pharyngeal gland lobe relatively long, well developed, ventrally overlapping of the intestine difficult to observe, three gland nuclei present. Hemizonid anterior, adjacent to the excretory pore.

Tail slightly curved ventrally, relatively long, gradually tapering until distinct short hyaline tail terminus, rectum usually inflated. Phasmids posterior to anus, small, located in ventral incisure of lateral field. One or two cuticular tail constrictions present.

Eggs (n=30)

Length 99-117 μm (106 ± 5.4); width 45-52 μm (48 ± 2.0); length / width ratio 2.0-2.5 (2.2 ± 0.2).

Hosts

Detected on *Elymus farctus* (Viv.) Meldris and *Ammophila arenaria* (L.) Link. Host suitability was studied in the greenhouse. *M. duytsi* failed to reproduce on *Lycopersicon esculentum* Mill. cv. Moneymaker, *Brassica oleracea laciniata* L. cv. Westlandse winterharde, *Zea mays* L. cv. Splenda and *Lolium multiflorum* Lamk., but reproduced on *Triticum aestivum* L. cv. Minaret.

Distribution

E. farctus is a clonal, salt resistant dune grass that colonises beaches. The plant grows in an extremely dynamic and unpredictable environment. Plants may be washed away during storms. When embryonal dunes are formed, *Ammophila arenaria* (marram grass) may colonise and, finally, replace *E. farctus*. *M. duytsi* has also been found in the root zone of *A. arenaria* near Oostvoorne, together with *M. maritima* Jepson, 1987.

A. arenaria is a native plant species from north-western Europe (Huiskes, 1979), and was used since the XVIII to stabilize dunes in the south-west coastal area of France. It occurs also in the Mediterranean and has been exported for sand stabilization to Australia, New Zealand, South Africa and the west coast of North-America.

As the hosts *E. farctus* and *A. arenaria* are very common beachgrasses close to the North Sea, it may be expected that *M. duytsi* is widely distributed in West-European coastal foredunes, para- or sympatrical with *M. maritima*.

To test this hypothesis we sampled during 1997 and 1998 on *E. farctus* and *A. arenaria* (see *M. maritima*) in coastal dunes of Belgium (Koksijde), Wales (Tywyn), England (Perranporth), France (Bray-Dune plage and Cabourg), Portugal (Dunas de Mira) and the Netherlands (several locations). In all the samples we observed *M. duytsi*. In the Netherlands this species is widely distributed along the North Sea foredunes, including all the northern islands.

Based on the present known distribution it is likely that *M. duytsi* also occurs in coastal dunes in Germany, Denmark and possibly Southern Norway and the West coast of Sweden.

Type locality and host

Collected and described from the roots of *Elymus farctus* (Viv.) Melderis (sand twitch), grown near the so called motorcar beach of the coastal foredunes near Oostvoorne (51°55' NL 5°06' EL), in a brackish environment.

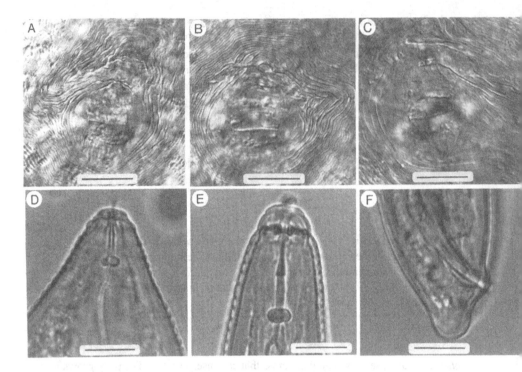

Fig. 2. *Meloidogyne duytsi*. A-D: Female. A-C: Perineal patterns; D: Head end (lateral); E-F: Male (lateral). E: Head end; F: Tail. (Scale bar A-C= 25 μm; D-F= 10 μm).

Type material

Holotype. Female on slide WT 3236, collection of Agricultural University, Wageningen the Netherlands. *Paratypes*. Two female perineal patterns and heads, two males and five J2's deposited at each of the following collections: Agricultural University, Wageningen the Netherlands (WT 3237-3239); University Gent, Zoology Institute, Gent Belgium; Rothamsted Experimental Station, Harpenden UK.

Etymology

The species name refers to Henk Duyts who found it near Oostvoorne for the first time and marked it as a species different from *M. maritima*.

Remarks

M. duytsi is characterized by a relatively large female body, slightly curved stylet of 13.3 μm (12.6-13.9) long with large transversely ovoid knobs, set off from the shaft. Perineal pattern asymmetrical with a low dorsal arch and coarse striae, and lateral wing present in most patterns. Male stylet length is 19.8 μm (19.0-20.2), knobs transversely ovoid, set off from the shaft. Head region high, labial disc slightly raised, lateral lips small, pointed tail. Second-stage juvenile body-, tail- and hyaline tail length 423.6 μm (403.2-454.4), 70.4 μm (65.1-76.5) and 11.3 μm (9.5-13.3) respectively. Hemizonid anterior and adjacent, to excretory pore. Tail curved ventrally.

M. duytsi induced no or only small galls on the hosts *E. farctus* and *A. arenaria*, therefore root infection is easily overlooked. A large egg-mass is produced by each female, protruding from the root. One or more males are present inside each egg-mass.

M. duytsi is also characterized by a unique malate dehydrogenase (Mdh) pattern, and one very slow esterase (Est) band. According to the enzyme phenotype coding of Esbenshade & Triantaphyllou (1985) the patterns are named N2 (Mdh) and VS1 (Est) respectively (Fig. 5 and Discussion).

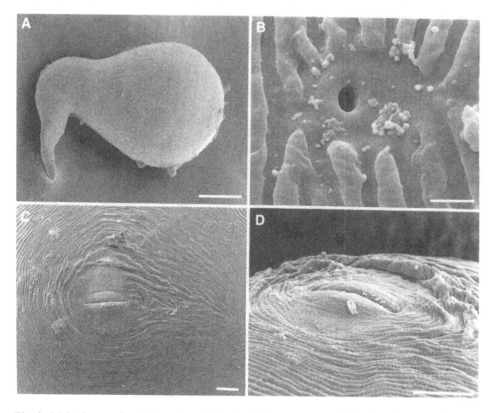

Fig. 3. *Meloidogyne duytsi*. Females. A: Body; B: Excretory pore; C: Perineal pattern; D: Perineal pattern (side view). (Scale bars: A-D, F= 1μm; E= 100 μm).

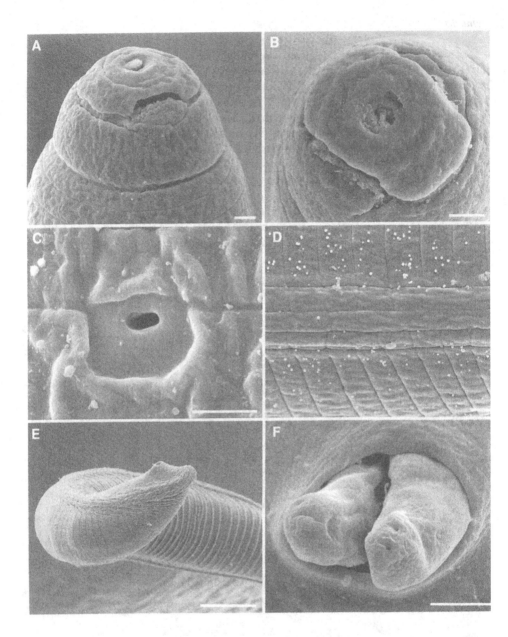

Fig. 4. *Meloidogyne duytsi*. Males. A: Cephalic region (lateral view); B: Cephalic region (face view); C: Excretory pore; D: Lateral field; E: Tail region (lateral view); F: Spicule. (Scale bar: A= 100 µm, B= 1 µm, C, D= 10 µm).

Table I

Morphometrics of Meloidogyne duytsi *Karssen, van Aelst & van der Putten, 1998*
[mean ± SD (range); n=30; all measurements in μm].

Character	Females	Males	J2
L	865±108.3	1316±173.6	424±13.7
	(560-960)	(960-1680)	(403-454)
Greatest body diam.	591±115.9	34.4±2.7	17.1±0.6
	(368-800)	(27.2-36.7)	(15.8-17.7)
Body diam. at stylet knobs	-	19.7±0.7	-
		(18.3-20.9)	
Body diam. at excr. pore	-	30.1±1.9	-
		(25.3-32.9)	
Body diam. at anus	-	-	12.3±0.5
			(11.4-13.3)
Head region height	-	5.4±0.5	2.4±0.3
		(4.4-6.3)	(1.9-3.1)
Head region diam.	-	11.1±0.5	6.0±0.3
		(10.1-12.0)	(5.7-6.3)
Neck length	271±52.2	-	-
	(192-336)		
Neck diam.	118±21.6	-	-
	(80-160)		
Stylet	13.3±0.3	19.8±0.3	11.1±0.4
	(12.6-13.9)	(19.0-20.2)	(10.7-12.0)
Stylet base-ant. end	-	-	15.1±0.6
			(13.3-15.8)
Stylet cone	-	10.3±0.3	-
		(9.5-10.8)	
Stylet shaft and knobs	-	9.5±10.1	5.9±0.3
		(8.2-10.1)	(5.7-6.3)
Stylet knob height	2.1±0.3	3.0±0.3	1.6±0.3
	(1.9-2.5)	(2.5-3.2)	(1.3-1.9)
Stylet knob width	4.7±0.4	5.3±0.3	2.7±0.4
	(4.4-5.1)	(5.1-5.7)	(1.9-3.2)
DGO	3.8±0.4	4.0±0.3	3.6±0.3
	(3.2-4.4)	(3.8-5.1)	(3.2-3.8)
Ant. end to metacorpus	-	74±5.0	55±2.0
		(63-87)	(49-59)
Metacorpus length	38.4±2.6	-	-
	(34.8-41.1)		
Metacorpus diam.	37.1±2.9	-	-
	(31.0-41.1)		
Metacorpus valve length	12.0±0.6	-	4.2±0.4
	(11.4-13.3)		(3.2-3.8)
Metacorpus valve width	9.6±0.8	-	3.8±0.2
	(8.9-12.0)		(3.2-4.4)
Excretory pore-ant. end	37.5±4.7	138±11.8	79±86.6
	(30.3-47.4)	(117-156)	(70-87)
Tail	-	13.2±1.5	70±3.1
		(10.1-15.8)	(65-77)
Tail terminus length	-	-	11.3±1.0
			(9.5-13.3)

Phasmids-post. end	-	8.6±1.1 (5.7-10.1)	-
Spicule	-	25.9±0.9 (24.0-27.2)	-
Gubernaculum	-	7.1±0.4 (6.3-7.6)	-
Testis	-	692±114.1 (379-853)	-
Vulva slit length	23.2±2.2 (19.6-25.3)	-	-
Vulva-anus distance	21.5±2.5 (15.8-26.5)	-	-
a	1.5±0.2 (1.2-1.8)	38.2±3.6 (30.3-45.8)	24.9±1.1 (23.2-27.3)
c	-	101±15.7 (76-131)	6.0±0.2 (5.6-6.6)
c'	-	-	5.7±0.3 (5.2-6.4)
T	-	53±6.6 (39-66)	-
Body length / neck length	3.3±0.7 (2.6-5.0)	-	-
Body length / ant. end to metacorpus valve	-	-	7.7±0.3 (7.2-8.6)
Stylet knob width / height	2.3±0.3 (1.8-2.7)	1.8±0.2 (1.6-2.3)	-
Metacorpus length / width	1.0±0.1 (1.0-1.1)	-	-
(Excretory pore / L) x 100	-	10.5±1.0 (8.5-13.5)	18.8±0.7 (16.3-19.9)

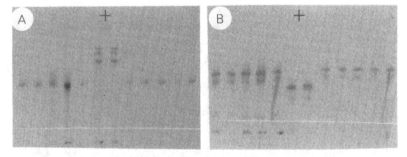

Fig. 5. *Meloidogyne duytsi*. Esterase (A) and malate dehydrogenase patterns (B) of the type population from the Netherlands (Oostvoorne); reference: *M. javanica* (China) in the two middle positions of the gel.

Meloidogyne graminis (Sledge & Golden, 1964) Whitehead, 1968

(Fig. 6 & 7)

Syn. *Hypsoperine graminis* Sledge & Golden, 1964
Hypsoperine (*Hypsoperine*) *graminis* (Sledge & Golden, 1964) Siddiqi, 1985

Meloidogyne graminis

Sledge & Golden (1964); *Proc. Helminth. Soc. Washington* 31: 83-88.
  Sledge (1962); *Plant Disease Reporter* 46: 52-54.
  Whitehead (1968); *Transactions of the Zoological Society of London* 31: 263-401.
  Grisham *et al.* (1974); *Phytopathology* 64: 1485-1489.
  Sturhan (1976); *Nachrichtenbl. Deut. Pflanzenschutzd.* 28: 113-117.
  Jepson (1987); *Identification of root-knot nematodes* (Meloidogyne *species*). Wallingford,
  UK: CAB International. 265 pp.

Measurements

(in glycerine) see Table II

Female

Body oval and asymmetrical shaped, relatively large with short neck. Perineal pattern located
on a posterior protuberance. Short stylet, cone slightly curved dorsally, knobs ovoid and
slightly backwardly sloping. S-E pore less than one stylet length behind head end. Perineal
pattern rounded to oval shaped, with coarse striae and angular dorsal arch, tail terminus area
free of striae, lateral field visible.

Male

Body anteriorly tapering. Head slightly set off from body. Head cap relatively small, labial
disc not elevated, lateral lips absent. Head region without transverse incisures. Short stylet
with rounded knobs, slightly backwardly sloping. DGO to stylet knobs relatively short.
Lateral field with four incisures, clear areolation not observed.

Second-stage juvenile

Relatively long and slender body. DGO close to stylet knobs. Hemizonid posterior to the S-E
pore. Long slender tail with inflated rectum and relatively long gradually tapering hyaline tail
part, ending in rounded terminus.

Hosts

Known to parasitize grasses, including different cultivars of both *Cynodon dactylon* (L.) Pers.
and *Zoysia japonica* Steud., *Paspulum notatum* Fluegge, *Stenotaphrum secundatum* (Walter)
Kuntze, *Oryza sativa* L., *Digitaria sanguinalis* (L.) Scop. and *Ammophila arenaria* (L.) Link.
(Jepson, 1987). Also detected on the cereals *Sorghum vulgare* Pers. and *Zea mays* L.

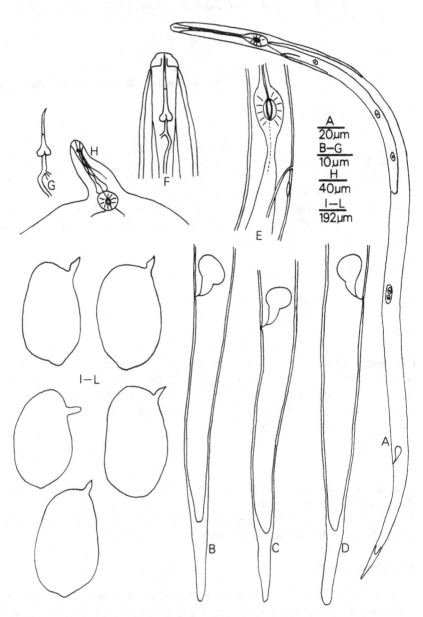

Fig. 6. *Meloidogyne graminis*. A-E: Second-stage juvenile (lateral). A: Body; B-D: Tails; E: Metacorpus region; F: Male head end (lateral); G-L: Female (lateral). G: Stylet; H: Anterior end; I-L: Body shapes (after Sledge & Golden, 1964).

Distribution

In Europe only reported from coastal dunes in Germany and the Netherlands.

Type locality & host

Lawn around Division of Plant Industry Laboratory, Winter Haven, Florida USA and described from *Stenotaphrum secundatum*.

Type material

Paratypes deposited at the USDA, Beltsville, Maryland USA were studied.

Etymology

The species epithet means 'from grass'.

Remarks

The studied slides are in agreement with the description. Male stylet knobs were described as set off by Jepson (1987), although slightly backwardly sloping knobs were observed. *M. graminis* was reported from the coastal dune grass *Ammophila arenaria,* in Germany and the Netherlands (Sturhan, 1976; Bongers, 1988). Jepson (1987) described *M. maritima* from the same host, originating from the UK. Recently, *M. maritima* was redescribed from the UK type locality, and additionally the presence of *M. maritima* on *A. arenaria* in the Netherlands was confirmed (Karssen *et al.*, 1998b). Also, study of slides with juveniles from coastal dunes on the island Sylt Germany, indicates the presence of *M. maritima* not *M. graminis*. Therefore we hereby state that *M. graminis* is not present in coastal dunes in Germany or the Netherlands. There is also no indication that it is present elsewhere in Europe.

Esbenshade & Triantaphyllou (1987) described the esterase (VS1 type) and malate dehydrogenase (N4 type) patterns for *M. graminis*.

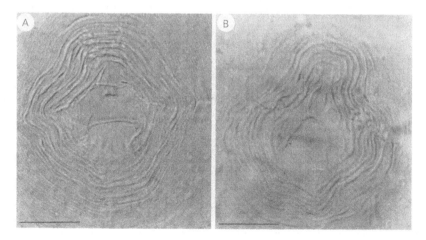

Fig. 7. *Meloidogyne graminis.* A & B: Female perineal patterns. Bar= 25 µm.

Table II

Morphometrics of Meloidogyne graminis (*Sledge & Golden, 1964*) *Whitehead, 1968*
[*mean ± SD (range); all measurements in μm*]

Character	J2	Males	Females
N	10	3	6
Body length	463±31.4 (403-506)	1248±96 (1152-1344)	-
Greatest body diameter	13.7±0.7 (12.6-15.2)	34.8±3.2 (31.6-37.9)	-
Body diam. at stylet knobs	-	14.8±0.8 (13.9-15.2)	-
,, ,, ,, S-E pore	12.4±0.4 (12.0-13.3)	25.0±1.0 (24.0-25.9)	-
,, ,, ,, anus	10.0±0.5 (9.5-10.7)	-	-
Stylet length	12.4±0.4 (12.0-13.4)	18.1±0.4 (17.7-18.3)	12.8±0.4 (12.6-13.5)
DGO	2.5±0.3 (1.9-3.2)	2.7±0.4 (2.5-3.2)	4.1±0.6 (3.3-5.0)
S-E pore to anterior end	-	-	11.4±1.6 (8.2-12.6)
Anterior end to metacorpus	-	61±4.9 (57-66)	99±7.3 (92-112)
Metacorpus length	-	-	36.0±5.0 (30.3-44.2)
,, diameter	-	-	35.3±3.1 (31.6-41.1)
Tail length	76±4.9 (66-82)	-	-
Tail terminus length	18.9±1.8 (15.8-21.5)	-	-
Anus-primordium	105±8.7 (89-116)	-	-
Spicule	-	27.4±3.6 (25.3-31.6)	-
Gubernaculum	-	8.0±0.4 (7.6-8.2)	-
a	33.9±1.9 (31.5-37.3)	36.0±0.5 (35.5-36.5)	-
b"	8.1±0.6 (6.9-8.9)	-	-
c	6.1±0.3 (5.7-6.4)	-	-
c'	7.7±0.5 (6.8-8.3)	-	-
T	-	66±11.8 (52-74)	-
(S-E pore/L)x 100	16.4±1.4 (15.2-19.4)	7.9±0.7 (7.3-8.7)	-

Meloidogyne kralli Jepson, 1983

(Fig. 8-11)

Meloidogyne kralli

Jepson (1983); *Revue Nématol.* 6: 239-245.

Measurements

(in glycerine) see Table III

Female

Body ovoid and asymmetrically shaped, with short neck. Perineal pattern located on a posterior protuberance. Short stylet, cone dorsally curved, knobs rounded and set off from shaft. S-E pore less than one stylet length behind head end. Perineal pattern rounded with relatively faint and fine striae, lateral field weakly visible. Anus not always covered with a cuticular fold.

Male

Rare. Body anteriorly tapering. Head not set off from body, relatively high, labial disc not elevated. Head region without transverse incisures. Stylet medium-sized, knobs robust and rounded, set off from shaft. Lateral field with 5 incisures, additional incomplete incisures visible, areolation not observed.

Second-stage juvenile

Body slender and moderately in length. Hemizonid anterior and adjacent to the S-E pore. Long slender tail with inflated rectum and relatively long gradually tapering hyaline tail part, ending in pointed tip. Lateral field with four incisures, sometimes five incisures visible.

Hosts

The following Cyperaceae hosts are parasitized in natural conditions: *Carex acuta* L., *C. vesicaria* L., *C. riparia* Curt., *C. pseudocyperus* L. and *Scripus sylvaticus* L. In the greenhouse it reproduced also on *Festuca pratensis* Huds. and *Hordeum vulgare* L. (Jepson, 1983). Although an attempt to maintain *M. kralli* from Estonia on *Triticum aestivum* L. and *Hordeum vulgare* L. failed, in the greenhouse at the Plant Protection Service.

Distribution

Detected in Estonia, Russia, Poland and the UK, all in wet sandy, peat and silt soils.

Type locality & host

Elva river near Elva, Tartu district Estonia, described from *Carex acuta.*

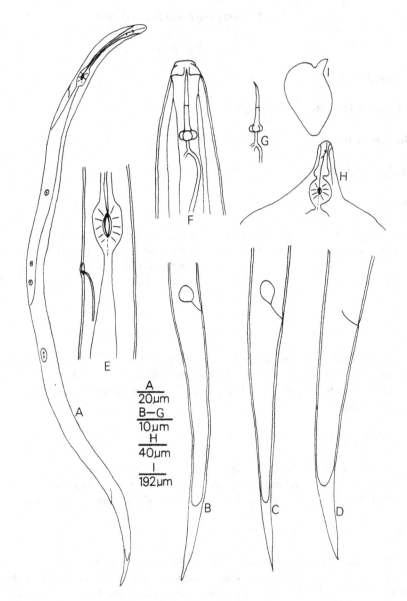

Fig. 8. *Meloidogyne kralli*. A-E: Second-stage juvenile (lateral). A: Body; B-D: Tails; E: Metacorpus region; F: Male head end (lateral); G-I: Female (lateral). G: Stylet; H: Anterior end; I: Body shape (after Jepson, 1983).

Type material

Paratypes deposited at Rothamsted Experimental Station, Harpenden UK, were studied.

Etymology

The species is named after Prof. dr. E. Krall, who found this species in 1968 for the first time.

Remarks

The studied paratypes are in agreement with the description, except for a smaller observed tail and hyaline tail length for J2's and rounded male and female stylet knobs versus rounded to transversely ovoid described knobs. The material for the description was based on lactophenol mounted specimens, collected in 1971. This material was not useful for scanning electron microscopy study (Jepson, 1983). Therefore Dr. E. Krall collected in 1997, on the type locality, some infected *Carex acuta* plants. This material was used for an additional SEM study (Fig. 10 & Karssen *et al.*, in prep.) and for isozyme electrophoresis. A N1c type malate dehydrogenase pattern (Fig. 11) and a very weak multiple banding esterase pattern were detected.

Judging from the known distribution, it is likely that *M. kralli* occurs also in natural habitats in Germany and the Netherlands.

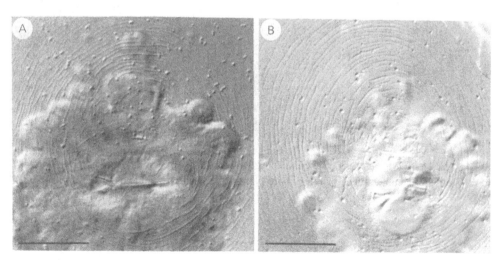

Fig. 9. *Meloidogyne kralli*. A & B: Female perineal patterns. Bar= 25 μm.

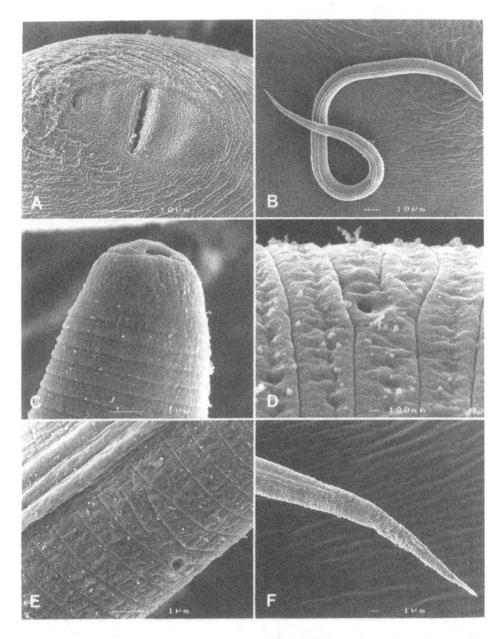

Fig. 10. *Meloidogyne kralli*. A: Female perineal pattern; B-F: Second-stage juvenile (lateral). B: Body; C: Head end; D: Excretory pore (ventral); E: Lateral field and anus (ventral); F: Tail.

Table III

Morphometrics of Meloidogyne kralli *Jepson, 1983*
[*mean ± SD (range); all measurements in µm*]

Character	J2	Males	Females
N	10	1	5
Body length	412±13.2 (394-432)	1042	-
Greatest body diameter	16.4±1.1 (15.2-18.3)	41.1	-
Body diam. at stylet knobs	-	15.2	-
„ „ „ S-E pore	13.8±0.9 (12.6-15.2)	30.3	-
„ „ „ anus	10.5±0.5 (10.1-11.4)	-	-
Stylet length	10.7±0.4 (10.1-11.4)	19.0	13.3±0.5 (12.6-13.9)
DGO	4.2±0.5 (3.8-5.1)	4.4	4.0±0.3 (3.8-4.4)
S-E pore to anterior end	-	-	16.7±2.7 (12.1-19.0)
Anterior end to metacorpus	-	75	58±7.9 (51-70)
Metacorpus length	-	-	35.0±1.4 (33.5-37.3
„ diameter	-	-	31.5±1.1 (29.7-32.9)
Tail length	61±2.1 (58-65)	-	-
Tail terminus length	15.4±1.4 (13.3-18.3)	-	-
Anus-primordium	110±5.4 (99-117)	-	-
Spicule	-	25.3	-
Gubernaculum	-	7.6	-
a	25.2±1.6 (23.1-28.2)	25.3	-
b"	7.4±0.5 (6.6-8.1)	-	-
c	6.8±0.3 (6.2-7.4)	-	-
c'	5.8±0.3 (5.3-6.3)	-	-
T	-	29.1	-
(S-E pore/L)x 100	16.1±1.5 (12.7-17.7)	12.1	-

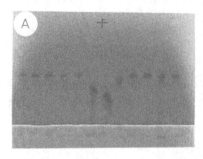

Fig. 11. *Meloidogyne kralli*. Malate dehydrogenase patterns of the type population from Estonia (Tartu); reference: *M. javanica* (China) in the two middle positions of the gel.

Meloidogyne maritima (Jepson, 1987) Karssen, van Aelst & Cook, 1998

(Fig. 12-16)

Meloidogyne maritima

Jepson (1987); *Identification of root-knot nematodes* (Meloidogyne *species*). Wallingford, UK: CAB International. 265 pp.
Karssen *et al.* (1998b); *Nematologica*. 44: 241-253.

Preface

In 1987 Jepson described *Meloidogyne maritima* from marram grass *Ammophila arenaria* (L.) Link, a sand-stabilizing coastal dune grass. This root-knot nematode species had been recorded from coastal dunes in Germany and the Netherlands (Sturhan, 1976; Brinkman, 1985) but was reported as *M. graminis* (Sledge & Golden, 1964) Whitehead, 1968.

During 1996, an unknown root-knot nematode was detected in the Dutch coastal foredunes on *Elymus farctus* (Viv.) Melderis, and described as *M. duytsi* Karssen, van Aelst & van der Putten, 1998a. Surprisingly, *M. duytsi* and *M. maritima* were found together on marram grass during supplementary sampling of several locations in Dutch and Belgian coastal foredunes.

Detailed morphological studies of the sampling material revealed that some of the *M. maritima* juvenile and male characteristics did not fit the description. A search for possible deviations in the type material was not possible, as types were not prepared and deposited in a nematode collection (Dr. D.J. Hooper and Dr. A. M. Golden, pers. comm.).

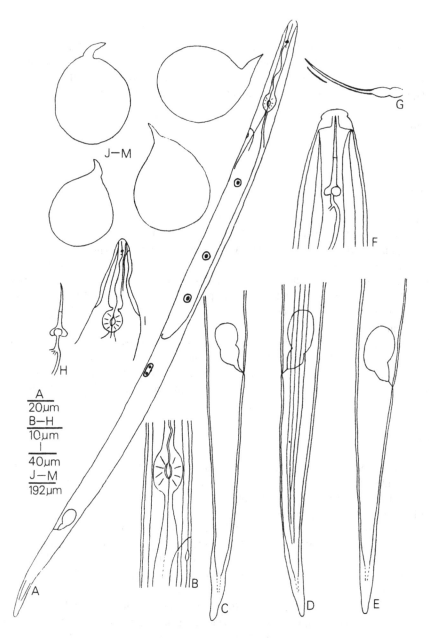

Fig. 12. *Meloidogyne maritima*. A-E: Second-stage juvenile. A: Body (lateral); B: Metacorpus region; C-E: Tails (lateral); F,G: Male. F: Head end (lateral); G: Spicule and gubernaculum (lateral); H-M: Female. H: Stylet (lateral); I: Anterior end (lateral); J-M: Body shape.

Meloidogyne maritima from the type locality at Perranporth UK did not differ in morphological and biochemical characteristics from the Dutch and Belgian populations. Also *M. duytsi* was found at the type locality. *Meloidogyne maritima* was recently redescribed from the type locality, a neotype and neoparatypes were prepared (Karssen *et al.*, 1998b).

Measurements

(in TAF) see Table IV

Female

Body globular shaped, annulated, pearly white, neck region relatively small but distinct, slightly projecting from body axis to one side, a posterior protuberance only rarely present. Head region not set off from body. Head cap distinct, variable in shape; labial disc slightly elevated. Cephalic framework distinct, weakly sclerotized; vestibule extension distinct.

Stylet cone curved dorsal; shaft cylindrical; knobs rounded to transversely ovoid, slightly sloping backwardly from the shaft. Excretory pore rounded, located near level of stylet knobs. Metacorpus rounded, without vesicles near lumen lining. Pharyngeal glands highly variable in shape and size.

Perineal pattern relatively small, rounded to weakly oval shaped; dorsal arch relatively low, with coarse striae. Tail terminus area distinct, without punctations. Lateral lines distinct, clearly separating dorsal and ventral pattern region. Phasmids small, inter-phasmidial distance less than mean vulva length. In most patterns anus covered with a fold. Ventral pattern region rounded, striae finer than in dorsal region.

Male

Body annulated (annule about 2 μm width), twisted, vermiform and anteriorly tapering. Lateral field with four incisures, middle and outer bands regularly areolated. Head not set off, single post-labial annule (head region) present, transverse incisures not observed. Head tapering anteriorly near stylet knobs level.

Labial disc oval, elevated and fused with medial lips. Hexagonal shaped prestoma present with six inner cephalic sensilla. Medial lips crescent shaped with small indention medially, four small slit-like cephalic sensilla observed. A relatively long slit-like amphidial opening present between labial disc and lateral side of post-labial annule. Lateral lips present, ranging from weak to clearly distinct.

Cephalic framework moderately sclerotized, vestibule extension distinct. Stylet cone straight, shaft cylindrical; knobs rounded, relatively small and slightly sloping posteriorly from shaft. Excretory pore rounded. Metacorpus oval, procorpus slender. Pharyngeal gland lobe overlapping intestine ventrally, two subventral gland nuclei present.

Hemizonid approximately 4 μm long and positioned anterior, or rarely posterior, to excretory pore. Testis variable in length, monorchic with outstretched, rarely reflexed, germinal zone. Tail rounded, short and twisted. Spicules slender and curved ventrally, two pores observed on each spicule tip. Phasmids located posterior to cloaca.

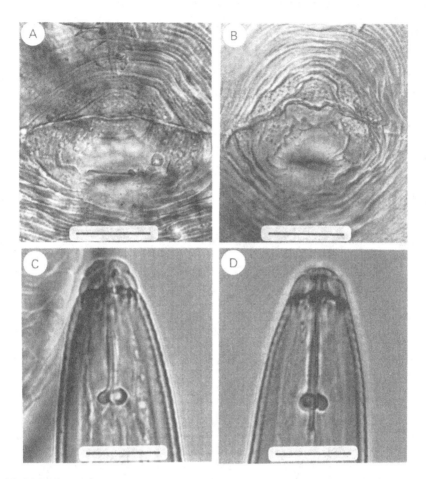

Fig. 13. *Meloidogyne maritima.* A, B: Female perineal patterns; C, D: Male head end (lateral, ventral). Bar= 25 μm in A, B; 10 μm in C, D.

Second-stage juveniles

Body vermiform, relatively long and annulated (annule about 1 μm width). Lateral field with four incisures, weakly areolated. Head region rounded to truncate, not set off from body. Cephalic framework weakly sclerotized, vestibule extension distinct. Slender stylet with straight cone and cylindrical shaft; knobs rounded and slightly sloping backwardly.

Metacorpus oval, triradiate lining weakly sclerotized, vesicles not observed on anterior lumen lining. Pharyngeal gland lobe variable in length, overlapping intestine ventrally, with three small gland nuclei. Hemizonid posterior not adjacent to excretory pore.

Tail relatively long, distinctly annulated, gradually tapering towards tail terminus end, anterior part of hyaline tail terminus usually indistinct, proctodeum inflated. Phasmids relatively small but distinct, posterior to anus near ventral incisure of lateral field. Tail tip usually slightly curved dorsally.

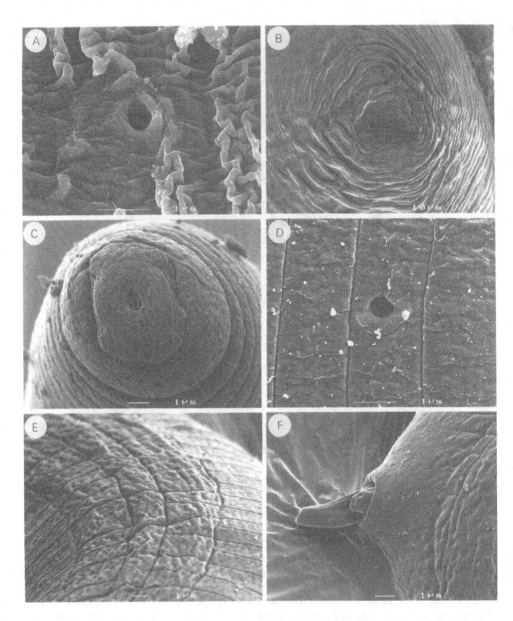

Fig. 14. *Meloidogyne maritima*. A-B: Female. A: Excretory pore; B: Perineal pattern; C-F: Male. C: *En face* view; D: Excretory pore; E: Lateral field; F: Cloacal region with spicule.

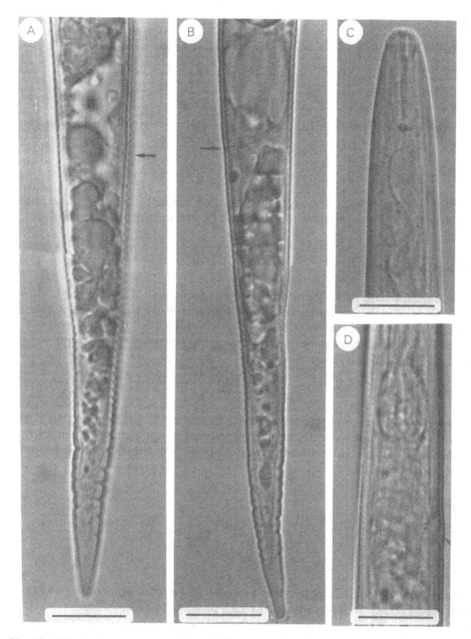

Fig. 15. *Meloidogyne maritima.* Second-stage juvenile. A-B: Tail shapes (lateral, arrow = anus); C: Head end; D: Metacorpus region (arrow = hemizonid). Bar = 10μm.

Eggs (n=30)

Length 99-122 μm (105 ± 5.8); width 42-50 μm (46 ± 2.4); length/ width ratio 2.1-2.7 (2.3 ± 0.2).

Hosts

Detected on *Ammophila arenaria* (L.) Link, *Elymus arenarius* L. and *Calamagrostis epigejos* (L.) Roth.

Distribution

Found in the coastal dunes in Germany (Sylt), Belgium (Koksijde), France (Bray-Dune plage and Cabourg), UK (Perranporth) and the Netherlands (several locations, including northern islands),

Type locality and host

Collected and redescribed from the roots of *Ammophila arenaria* (L.) Link, growing in the seaward slopes of mobile calcareous sanddunes at about 20m above mean sea level and about 1 km north of Perranporth, Cornwall UK (O.S. Map reference SX 760 560). Only roots attached to *A. arenaria* rhizomes were collected from areas dominated by *A. arenaria*, with occasional *Festuca* and *Carex* spp.

Type material

Neotype female, Rothamsted Experimental Station, Harpenden UK. *Neoparatypes*: one female perineal pattern and head, two males and four J2s deposited each at: Agricultural University, Wageningen the Netherlands; University Gent, Zoology Institute, Gent Belgium; Rothamsted Experimental Station, Harpenden UK.

Etymology

The species name means 'from or at the sea'.

Remarks

Meloidogyne maritima is characterized by a female with a stylet of 14.2 μm (13.9-15.2) length, curved dorsally with rounded to transversely ovoid knobs, slightly sloping backwardly; excretory pore relatively close to the head end; small, rounded perineal pattern with distinct lateral lines; by a male with a stylet of 20.5 μm (19.6-22.1) length with relatively small rounded knobs, slightly sloping backward; head not set off, tapering, lateral lips present, rounded labial disc fused with crescent shaped medial lips; by a second-stage juvenile body length of 471 μm (441-512), hemizonid posterior to the excretory pore and an indistinct hyaline tail terminus.

Additionally, young females have a malate dehydrogenase isozyme pattern N1c type, and an esterase VS1-S1 pattern.

M. duytsi differs from *M. maritima* by a shorter juvenile body length, an anterior hemizonid position and a distinct hyaline tail terminus, a non-tapering male head with large transversely ovoid knobs, set off from the shaft and a longer DGO, a perineal pattern without distinct lateral lines and different malate dehydrogenase and esterase patterns.

Morphologically and ecologically related species of the 'graminis-group', like *M. graminis*, *M. kralli* Jepson, 1983, *M. naasi* Franklin, 1965 and *M. sasseri* Handoo, Huettel & Golden, 1993 differ in morphology and morphometrics from the redescribed *M. maritima* (Discussion and Karssen *et al.*, 1998a).

Jepson (1987) described in detail the relationship of *M. maritima* with the above mentioned 'graminis-group', including *M. aquatilis* Ebsary & Eveleigh, 1983, *M. graminicola* Golden & Birchfield, 1965, *M. oryzae* Maas, Sanders & Dede, 1978, *M. ottersoni* (Thorne, 1969) Franklin, 1971, *M. sewelli* Mulvey & Anderson, 1980 and *M. spartinae* (Rau & Fassuliotis, 1965) Whitehead, 1968.

Species occurring in the same geographical region, like *M. ardenensis* Santos, 1968, *M. artiellia* Franklin, 1968, *M. chitwoodi* Golden *et al.*, 1980 and *M. hapla* Chitwood, 1949 were compared by Karssen (1996) and in chapter 5 (see *M. fallax*).

Table IV

Morphometrics of Meloidogyne maritima *Jepson, 1987*
[*mean ± SD (range); n=30; all measurements in μm*]

Character	Females	Males	J2
L	748±136	1034±157	471±21.3
	(496-896)	(749-1357)	(442-512)
Greatest body diam.	530±88.7	31.1±11.1	17.0±0.7
	(352-624)	(24.0-35.4)	(15.8-18.3)
Body diam. at stylet knobs	-	17.7±1.5	-
		(13.9-19.6)	
Body diam. at excr. pore	-	25.8±3.5	15.0±0.4
		(18.3-31.6)	(14.5-15.8)
Body diam. at anus	-	-	12.4±0.6
			(11.4-13.9)
Head region height	-	5.1±0.8	2.6±0.4
		(4.4-7.0)	(1.9-3.2)
Head region diam.	-	9.8±0.5	5.9±0.3
		(8.9-10.1)	(5.7-6.2)
Neck length	179±55.5	-	-
	(128-304)		
Neck diam.	75±16.5	-	-
	(48-112)		
Stylet	14.2±0.5	20.5±0.7	12.4±0.3
	(13.9-15.2)	(19.6-22.1)	(12.0-12.6)

Stylet base-ant. end	-	-	16.7±0.7 (15.8-17.7)
Stylet cone	-	10.9±0.5 (9.5-11.4)	
Stylet shaft and knobs	-	9.6±0.6 (8.9-10.7)	6.1±0.3 (5.7-6.3)
Stylet knobs height	2.0±0.2 (1.9-2.5)	2.6±0.3 (2.5-3.2)	1.6±0.3 (1.3-1.9)
Stylet knobs width	4.7±0.4 (4.4-5.1)	5.1±0.3 (4.4-5.7)	2.9±0.4 (2.5-3.2)
DGO	3.6±0.3 (3.2-3.8)	2.8±0.4 (2.5-3.2)	2.9±0.3 (2.5-3.2)
Ant. end to metacorpus	-	78±3.8 (71-82)	65±3.2 (59-73)
Metacorpus length	30.7±3.3 (26.5-37.9)	-	-
Metacorpus diam.	28.5±4.6 (22.8-37.9)	-	-
Metacorpus valve length	11.3±0.4 (10.7-12.0)	-	3.7±0.2 (3.2-3.8)
Metacorpus valve width	9.1±1.8 (5.1-12.0)	-	3.1±0.2 (2.5-3.2)
Excretory pore-ant.end	15.0±2.9 (9.5-18.9)	109±12.6 (90-139)	81±3.0 (75-87)
Tail	-	12.1±0.7 (10.7-12.6)	72±2.5 (66-76)
Phasmids-post. end	-	6.8±1.4 (5.1-9.5)	-
Spicules	-	28.9±0.5 (27.8-29.7)	
Gubernaculum	-	7.3±0.5 (6.3-7.6)	-
Testis	-	501±95 (256-723)	-
Vulva slit length	22.2±1.5 (19.2-24.0)	-	-
Vulva-anus distance	17.9±2.1 (16.0-22.4)	-	-
a	1.4±0.1 (1.3-1.7)	33.3±3.0 (27.6-38.4)	27.8±1.6 (25.8-31.0)
c	-	86±14.4 (66-109)	6.6±0.2 (6.2-7.0)
c'	-	-	5.8±0.3 (5.2-6.3)
T	-	49±9.5 (26-67)	-
Body length / neck length	4.4±1.1 (2.9-6.5)	-	-
Body length / ant. end to metacorpus valve	-	-	7.3±0.3 (6.8-7.7)
(excretory pore / L) x 100	-	10.7±1.3 (8.4-12.9)	17.2±0.5 (16.6-18.6)

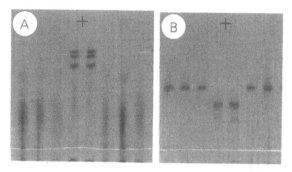

Fig. 16. *Meloidogyne maritima*. Esterase (A) and malate dehydrogenase patterns (B) of the type population from the UK (Perranporth); reference: *M. javanica* (China) in the two middle positions of the gel.

Meloidogyne naasi Franklin, 1965

(Fig. 17-19)

Meloidogyne naasi

Franklin (1965); *Nematologica* 11: 79-86.
  Whitehead (1968); *Transactions of the Zoological Society of London* 31: 263-401.
  Franklin (1973); *Meloidogyne naasi. C.I.H. Descriptions.* Set 2, No. 19. CAB, St. Albans, U.K.
  Sturhan (1976); *Nachrichtenbl. Deut. Pflanzenschutzd.* 28: 113-117.
  Jepson (1987); *Identification of root-knot nematodes* (Meloidogyne *species*). Wallingford, UK: CAB International. 265 pp.
  Eisenback & Hirschmann (1991); In: *Manual of agricultural nematology.* Ed. W.R. Nickle. New York: Marcel Dekker. pp: 191-274.

Measurements

(in glycerine) see Table V

Female

Body rounded, neck variable in length, with slight posterior protuberance. Stylet relatively short, cone slightly curved dorsally, knobs ovoid and backwardly sloping. S-E pore near stylet knobs level. Perineal pattern rounded to oval, phasmids prominent, striae relatively coarse, tail terminus area free of striae, lateral field not visible.

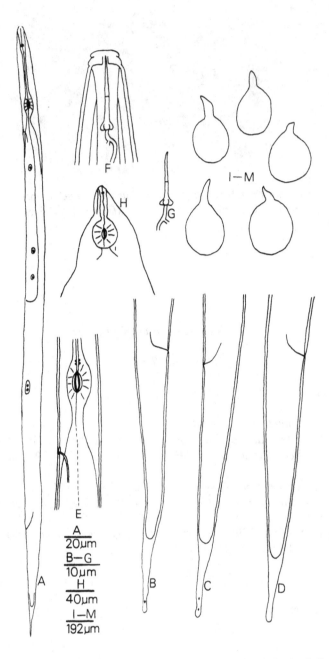

Fig. 17. *Meloidogyne naasi*. A-E: Second-stage juvenile (lateral). A: Body; B-D: Tails; E: Metacorpus region; F: Male head end (lateral); G-M: Female (lateral). G: Stylet; H: Anterior end; I-M: Body shape (after Franklin, 1965).

Male

Body anteriorly tapering. Head set off from body, labial disc not elevated, lateral lips absent. Head region without transverse incisures. Stylet short, knobs ovoid and backwardly sloping. DGO to stylet knobs relatively short. Several small vesicles present in the anterior lumen lining of the metacorpus. Lateral field with four incisures, areolated.

Second-stage juvenile

Body slender of moderately length. Anterior metacorpus lumen lining with several small vesicles. Hemizonid anterior and adjacent to the S-E pore. Tail relatively long and slender, rectum not inflated, hyaline tail part long and slender, ending in fine rounded tip.

Hosts

Several times reported on different grasses and cereals (Franklin, 1973; Jepson, 1987). Incidentally reported from a few dicotyledonous plants like sugar beet and weeds, but they are regarded as not very good hosts. Galls on grasses are relatively small, sometimes terminal, hooked or with lateral roots. The females are very often completely embedded within the root tissue.

Distribution

Widely distributed in Europe, reported from Wales, England, the Netherlands, Germany, Russia, Belgium, France, Italy and former Yugoslavia.

Type locality & host

Pendeck Farm, Tytherington, Gloucestershire UK, described from *Hordeum vulgare* L.

Type material

Paratypes deposited at Rothamsted Experimental Station, Harpenden UK were studied.

Etymology

The species epithet is based on the abbreviation NAAS, i.e. the National Agricultural Advisory Service UK.

Remarks

The studied material is in agreement with the description, except for ovoid observed male and female stylet knobs versus rounded described knobs. The vesicles in the male and second-stage juvenile metacorpus are not or only slightly visible in permanent slides. However in non- or fresly fixed juveniles, they are observed easily. These vesicles are not exclusively found in *M. naasi*, but also observed in *M. sasseri* juveniles (Karssen, 1996).

Esbenshade & Triantaphyllou (1985) described an esterase VF1 type and a malate dehydrogenase N1a type for *M. naasi* originating from the UK; populations from the Netherlands and Belgium are in agreement with these types.

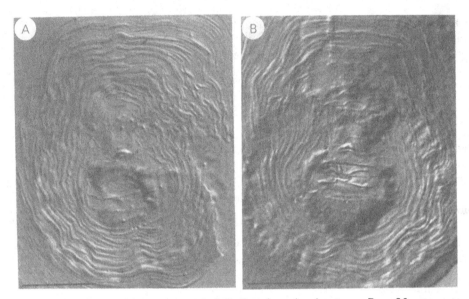

Fig. 18. *Meloidogyne naasi*. A & B: Female perineal patterns. Bar= 25 μm.

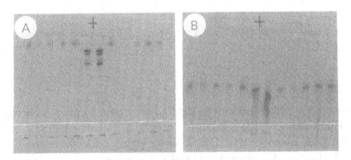

Fig. 19. *Meloidogyne naasi*. Esterase (A) and malate dehydrogenase patterns (B) of populations from the Netherlands (Wageningen and Kampen) and Belgium (Hasselt); reference: *M. javanica* (China) in the two middle positions.

Table V

Morphometrics of Meloidogyne naasi *Franklin, 1965*
[mean ± SD (range); all measurements in µm]

Character	J2	Males	Females
N	4	5	8
Body length	421±8.1	966±167	-
	(410-429)	(742-1088)	
Greatest body diameter	14.1±0.6	29.6±7.0	-
	(13.3-14.5)	(20.2-34.8)	
Body diam. at stylet knobs	-	12.6±1.5	-
		(10.7-13.9)	
„ „ „ S-E pore	12.3±0.4	20.4±3.5	-
	(12.0-12.6)	(15.8-23.4)	
„ „ „ anus	9.8±0.4	-	-
	(9.5-10.1)		
Stylet length	13.3±0.5	17.8±0.8	13.5±0.3
	(12.6-13.9)	(16.4-18.3)	(13.3-13.9)
DGO	2.9±0.4	2.9±0.4	3.7±0.4
	(2.5-3.2)	(2.5-3.2)	(3.2-4.4)
S-E pore to anterior end	-	-	13.4±0.9
			(12.6-14.5)
Anterior end to metacorpus	-	60±2.1	55±2.9
		(57-63)	(48-57)
Metacorpus length	-	-	39.4±3.3
			(34.1-44.2)
„ diameter	-	-	37.3±2.6
			(31.6-39.2)
Tail length	66±3.9	-	-
	(61-70)		
Tail terminus length	17.9±1.8	-	-
	(15.8-19.6)		
Anus-primordium	96±1.7	-	-
	(95-98)		
Spicule	-	28.3±1.4	-
		(27.2-30.3)	
Gubernaculum	-	7.5±0.5	-
		(7.0-8.2)	
a	30.0±1.5	33.2±2.5	-
	(28.3-31.8)	(31.1-36.8)	
b"	8.0±0.1	-	-
	(7.9-8.1)		
c	6.4±0.3	-	-
	(6.0-6.7)		
c'	6.8±0.5	-	-
	(6.1-7.4)		
T	-	61±14.1	-
		(49-84)	
(S-E pore/L)x 100	*17.4±0.3	11.6±1.9	-
	(17.0-17.7)	(10.0-14.7)	

Discussion

M. duytsi

For *M. duytsi*, the isozyme malate dehydrogenase (Mdh) N2 type was described (Karssen *et al.*, 1998a). Within the genus *Meloidogyne* a malate dehydrogenase pattern of one band is most common, two bands are rather unusual for a dimeric enzyme (Esbenshade & Triantaphyllou, 1985 & 1987). A critical look at Fig. 5B suggests it is a N1c type with an additional weak band or smear, a type of pattern very often caused by salt disturbance. The isolated *M. duytsi* females however always produced this type of Mdh pattern, even after prolonged desalting.

During the isolation process, young egg-laying females are first placed in an isotonic solution (0.9 % NaCl). Before sample preparation they are transferred to reagent-grade water for a few minutes to desalt (Karssen *et al.*, 1995). Normally *Meloidogyne* females will burst after a few minutes if not placed in an isotonic solution, however *M. duytsi* did not burst when directly placed in pure water. *M. duytsi* was described from a brackish environment, apparently it is highly tolerant and possibly adapted to quickly regulate it's internal pressure.

Two unusual small slit-like openings were observed in the male head, each present between the prestoma and amphidial opening (Fig. 4A & B). Although observed in all studied *M. duytsi* males, they are not clearly present in other known *Meloidogyne* male species. However in the male *en face* views of *M. nataliei* and *M. microcephala* they are also weakly present (Jepson, 1987). There is also no indication that these slit-like openings are preparation artefacts. The slit size is comparable with female and second-stage juvenile amphids. Under unfavorable conditions female juveniles develop into males with one or two gonads of variable length, a process known as sex reversal (Papadopoulou & Triantaphyllou, 1982). Probably the slit-like openings are related to J2 amphids or remnants of sex reversed female amphids.

M. maritima

The original *M. maritima* female description is in general agreement with the redescription, except for the observed stylet length, 13.5 µm (n=1) versus 14.2 µm (13.9-15.2) (n=30), and stylet knob position set off versus slightly sloping backwardly. In view of the fact that only a small number of females (n≤ 5) were measured by Jepson (1987), these differences are not significant.

The discrepancies between the male and juvenile data and the original description is more serious. The male lateral field was described as not areolated; the light and scanning electron microscopical obervations clearly show four lateral lines with areolated outer and inner bands (Fig. 14E). The male stylet knobs were described as '*offset and projecting anteriorly to make an acute angle with shaft, each knob smooth, transversely ovoid with the anterior surface concave*' versus small rounded knobs, slightly sloping backwardly. Concave knobs are rather unusual within the genus *Meloidogyne* Göldi, 1892, but are common in the subfamily Heteroderinae Filipjev & Schuurmans Stekhoven, 1941.

Possibly the males of *M. maritima* were mixed with males of *Heterodera arenaria* (Cooper, 1955) Robinson *et al.*, 1996. Jepson (1987) noted already that the stylet knobs she described '*are more typical of genera in the Heteroderidae*'. The male cyst nematode has concave stylet knobs and *H. arenaria* is also a parasite of *Ammophila arenaria* being widespread in coastal dunes in the Netherlands and the UK (de Rooij-van der Goes *et al.*, 1995; Robinson *et al.*, 1996). Cysts of *H. arenaria* were present in the samples from Perranporth from which Dr. R. Cook collected the examined *M. maritima* specimens.

The morphometrical ranges in juvenile body length (388-486 µm) and hyaline tail length (9.0-17.1 µm) in the original description of *M. maritima* are large as compared with the redescription (Table IV), whereas the original tail length (59.4-68.4 µm) is almost out of the range of this data. Finally the observed indistinct hyaline tail part was originally described as a '*delimiting end portion*'.

The status of the original holotype data, a second-stage juvenile with eight morphometrical characters, is unclear. The body and tail terminus lengths are closer to *M. duytsi* but others, like DGO, fit better with *M. maritima*. A reasonable explanation is that the original *M. maritima* juvenile data is based on two species: *M. maritima* and *M. duytsi*.

These explanations cannot be confirmed since no type material is available and specimens on original slides in the Rothamsted collection have dissolved through being mounted in TAF. Therefore Dr. R. Cook resampled in the dune system near Perranporth, which is the only available information on the type locality. Additional samples from the foreshore dunes at this place, where *E. farctus* and *A. arenaria* grew together proved to have a mixture of both *M. duytsi* and *M. maritima*, and in dunes with pure *A. arenaria*, *H. arenaria* was present. It is therefore concluded that the original description was based upon mixed material.

M. naasi

In the autumn of 1995, together with Mr. Bert Schoemaker, we sampled near the river Rhine and detected an unusual *M. naasi* population on different grass species at one location close to Wageningen, the Netherlands. As mentioned before, second-stage juveniles of *M. naasi* have a number of vesicles in the anterior lumen lining of the metacorpus, a useful character for routine identification (Franklin, 1973).

Surprisingly these vesicles were only noticed in about 50% of the observed number of juveniles, the others were completely free of these vesicles. An additional isozyme and morphological study confirmed that it was indeed *M. naasi*. Although there is no information available about the true nature of the vesicles, it is apparently not a stable taxonomical character for this population. Even after culturing this population on *Triticum aestivum* L. for more than 2 years, did not change the vesicle composition and ratio, as compared to the first observations.

Based on this instability it is hypothesized that these vesicles are not a local fold or bulge out of the metacorpial lumen lining, but a so far unknown physiological effect possibly caused by the dorsal pharyngeal gland granules (granuler enclosers?).

Comparable vesicles were also described for *Aphelenchoides megadorus* Allen, 1941.

'Graminis-group'

M. duytsi, *M. kralli*, *M. maritima* and *M. naasi* all belong to the 'Graminis-group' (see introduction). It is not only a group of species restricted to monocotyledons, but they are also morphological and morphometrical very similar. Therefore these species must be compared with great care.

The most distinct differences between the European members of the 'Graminis-group' are summerized in table VI. *Meloidogyne sasseri* Handoo *et al.*, 1993 is included, as it is involved in the degeneration of the North American beach-grass *Ammophila breviligulata* Fern.

(Seliskar & Huettel, 1993). Also *M. graminis* is included to illustrate the close morphological relationship of this (former European) species. All these species have one common morphological character: relatively long and slender second-stage juvenile tails.

Table VI

Characteristics of some members of the *Meloidogyne* 'graminis-group'
(all measurements in µm).

Character/species	M. duytsi	M. maritima	M. kralli	M. naasi	M. graminis	M. sasseri*
n	30	30	5	8	6	60
stylet length	13.3	14.2	13.3	13.5	12.8	14
	(12.6-13.9)	(13.9-15.2)	(12.6-13.9)	(13.3-13.9)	(12.6-13.5)	(13-15)
♀ knob shape	transversely ovoid	rounded to ovoid	rounded	rounded-pear shape	transversely ovoid	rounded
posterior pro tuberance-	no	rare	yes	slight	yes	slight
n	30	30	1	5	3	45
stylet length	19.8	20.5	19.0	17.8	18.1	20.0
	(19.0-20.2)	(19.6-22.1)		(16.4-18.3)	(17.7-18.3)	(19-21.5)
♂ knob shape	transversely ovoid	rounded	rounded	pear shape	rounded	rounded
knob position	set off	slightly sloping	set off	posteriorly sloping	slightly sloping	posteriorly sloping
DGO	4.0	2.8	4.4	2.9	2.7	?
n	30	30	10	4	10	50
body length	424	471	412	421	463	542
	(403-454)	(442-512)	(394-432)	(410-429)	(403-506)	(490-575)
tail „	70	72	61	66	76	97
	(65-77)	(66-76)	(58-65)	(61-70)	(66-82)	(85-107)
J2 hyaline tail	11.3	indistinct	15.4	17.9	18.9	19.9
	(9.5-13.3)		(13.3-18.3)	(15.8-19.6)	(15.8-21.5)	(16.0-23.0)
hemizonid rel. to excr. pore	anterior	posterior	anterior	anterior	posterior	posterior
vesicles in metacorpus	no	no	no	yes	no	yes

* Data from the original species description.

For the following reasons is the 'graminis-group' (*sensu* Jepson, 1987), as defined in the introduction, not correct: 1) The European species are indeed restricted to Poaceae and Cyperaceae host. 2) However the female body is not in all species markedly elongate, like *M. naasi*. 3) Only *M. kralli* females have their vulva situated on a posterior protuberance. 4) And are together with *M. duytsi* confined to damp or wet conditions. 5) Galling of monocotyledon roots is indeed slight or absent, but this is a general statement for root-knot nematodes on monocotyledons. For example *M. fallax* (chapter 5) induces on grasses also small galls, compared to galls induced on dicotyledons. 6) Key morphological characters for the group have indeed dimensions within narrow limits. 7) Only *M. duytsi* and *M. maritima* males are common in field soil. 8) The mode of reproduction is unknown for *M. duytsi*, *M. kralli* and *M. maritima*; facultative meiotic parthenogenesis is not limited to the 'graminis-group'.

Therefore the following redefinition of the 'graminis-group' is proposed: A) In the field they are confined to Poaceae and/or Cyperaceae hosts. B) Key morphological characters for the group have dimensions within narrow limits. C) Second-stage juveniles have relatively long and slender tails.

References

BONGERS, T. (1988). *De nematoden van Nederland*. Utrecht, K.N.N.V. 408 pp.

BRINKMAN, H. (1985). Planteparasitaire aaltjes bij helmgras (*Ammophila arenaria*). *Versl. en Meded. PlZiektenKd. Dienst Wageningen* 164, 73.

COURTNEY, W.D., POLLEY, D. & MILLER, V.L. (1955). TAF, an improved fixative in nematode technique. *Pl. Dis. Reptr.* 39, 570-571.

DE ROOIJ-VAN DER GOES, P.C.E. M. (1995). The role of plant-parasitic nematodes and soil-borne fungi in the decline of *Ammophila arenaria* (L..) Link. *New Phytol.* 129, 661-669.

DE ROOIJ-VAN DER GOES, P.C.E. M., VAN DER PUTTEN, W.H. & VAN DIJK, C. (1995). Analysis of nematodes and soil-borne fungi from *Ammophila arenaria* (Marram grass) in Dutch coastal foredunes by multivariate techniques. *Euro. Plant Pathol.* 101, 149-162.

ESBENSHADE, P.R. & TRIANTAPHYLLOU, A.C. (1985). Use of enzyme phenotypes for identification of *Meloidogyne* species. *J. Nematol.* 17, 6-20.

ESBENSHADE, P.R. & TRIANTAPHYLLOU, A.C. (1987). Enzymatic relationship and evolution in the genus *Meloidogyne* (Nematoda: Tylenchida). *J. Nematol.* 19, 8-18.

FRANKLIN, M.T. (1973). *Meloidogyne naasi. C.I.H. Descriptions of plant-parasitic nematodes*. Set 2, No. 19, St. Albans, UK, C.A.B.

GÖLDI, E.A. (1892). Relatoria sôbre a molestia do cafeiro na provincia da Rio de Janeiro. *Archos Mus. nac.,Rio de Janeiro* 8, 7-112.

HUISKES, A.H.L. (1979). Biological flora of the British isles: *Ammophila arenaria* (L.) Link [*Psamma arenaria* (L.) Roem. et Schult.: *Calamagrostis arenaria* (L.) Roth]. *J. Ecol.* 67, 363-382.

JEPSON, S.B. (1983). *Meloidogyne kralli* n. sp. (Nematoda: Meloidogynidae), a root-knot nematode parasitizing sedge (*Carex acuta* L.). *Rev. Nématol.* 6, 291-309.

JEPSON, S.B. (1987). *Identification of root-knot nematodes* (Meloidogyne *species*). Wallingford, UK, C.A.B. International. 265 pp.

KARSSEN, G., VAN HOENSELAAR, T., VERKERK-BAKKER, B. & JANSSEN, R. (1995). Species identification of root-knot nematodes from potato by electrophoresis of individual females. *Electrophoresis* 16, 105-109.

KARSSEN, G. (1996). On the morphology of *Meloidogyne sasseri* Handoo, Huettel & Golden, 1993. *Nematologica* 42, 262-264.

KARSSEN, G., VAN AELST, A. & VAN DER PUTTEN, W.H. (1998a). *Meloidogyne duytsi* n. sp. (Nematoda: Heteroderidae), a root-knot nematode from Dutch coastal foredunes. *Fundam. Appl. Nematol.* 21, 299-306.

KARSSEN, G., VAN AELST, A. & COOK, R. (1998b). Redescription of the root-knot nematode *Meloidogyne maritima* Jepson, 1987 (Nematoda: Heteroderidae), a parasite of *Ammophila arenaria* (L.) Link. *Nematologica* 44, 241-253.

LITTLE, L.R. & MAUN, M.A. (1996). The '*Ammophila* problem' revisited: a role for mycorrhizal fungi. *J. Ecol.* 84, 1-7.

MAAS, P.W.T.H., BRINKMAN, H. & PAVLICKOVA, E. (1987). Meloidogynae: wortelknobbelaaltjes. *Versl. en Meded. PlZziektenKd. Dienst Wageningen.* 166, 115-116.

PAPADOPOULOU, J. & TRIANTAPHYLLOU, A.C. (1982). Sex differentiation in *Meloidogyne incognita* and anatomical evidence of sex reversal. *J. Nematol.* 14, 549-566.

ROBINSON, A.J., STONE, A.R., HOOPER, D.J. & ROWE, J.A. (1996). A redescription of *Heterodera arenaria* Cooper, 1955, a cyst nematode from marram grass. *Fundam. Appl. Nematol.* 19, 109-117.

SEINHORST, J.W. (1959). A rapid method for the transfer of nematodes from fixative to anhydrous glycerin. *Nematologica* 4, 67-69.

SELISKAR, D.M. & HEUTTEL, R.N. (1993). Nematode involvement in the die out of *Ammophila breviligulata* (Poaceae) on the Mid-Atlantic coastal dunes of the United States. *J. coast. Res.* 9, 97-103.

SLEDGE, E.B. & GOLDEN, A.M. (1964). *Hypsoperine graminis* (Nematoda: Heteroderidae) a new genus and species of plant parasitic nematode. *Proc. Helminthol. Soc. Wash.* 31, 83-88.

STURHAN, D. (1976). Freilandvorkommen von *Meloidogyne*-Arten in der Bundesrepublik Deutschland *Nachrichtenbl. Deut. Pflanzenschutzd.* 28, 113-117.

VAN DER PUTTEN, W.H., VAN DIJK, C. & PETERS, B.A.M. (1993). Plant-specific soil-borne diseases contribute to succesion in foredune vegetation. *Nature* 362, 53-55.

WERGIN, W.P. (1981). Scanning electron microscopic techniques and applications for use in nematology. In: *Plant parasitic nematodes, Vol. 3*. pp. 175-204. Eds. B.M. Zuckerman & R.A. Rohde R.A. London, Academic Press.

5

REVISION OF THE

EUROPEAN ROOT-KNOT NEMATODES

III. ON MONO- AND DICOTYLEDONS

'J.H. Schuurmans Stekhoven fragt ob es möglich ist festzustellen,
welche von den von Chitwood und anderen beschriebenen Arten mit
der von Cornu als *marioni* beschrieben Art identisch ist. Wenn das
Möglich sein würde, denn würde der Diskussionsredner es bevorzugen,
Meloidogyne marioni für diese Art wiederherzustellen.'

In: M.T. Franklin (1957).

Introduction

This is the final part of the revision of the European root-knot nematodes, and comprises species parasitizing on mono- and dicotyledonous field hosts. It is the largest species group within the European root-knot nematodes, and includes the most successful species of the genus *Meloidogyne*. The majority of them are known as important agricultural pests in Europe and other regions of the world (Eisenback & Hirschmann, 1991). Recently two members of this group, *M. chitwoodi* and *M. fallax*, have been added to the European list of quarantine organisms, to prevent further spread within Europe.

The noticed Chitwood (1949) species problem (chapter 2) needs serious attention. Although this nomenclatorial problem was discussed in detail by Gillard (1961) and partly solved by Whitehead (1968), the confusion still exist. Therefore the status of the Chitwood (1949) species is discussed again.

Material and Methods

The group includes the following studied species: *M. arenaria*, *M. artiellia*, *M. chitwoodi*, *M. fallax*, *M. hispanica*, *M. incognita*, *M. javanica* and *M. kirjanovae*. Type specimens and isozyms were studied as described in chapter 3.

The added extensive description of *M. fallax* Karssen, 1996 was based on the same procedure as described for *M. duytsi* and *M. maritima* (chapter 4). For scanning electron microscopical observations however, a Jeol JSM 5200 operating at 15 kV accelerating voltage was used.

Species descriptions

Meloidogyne arenaria (Neal, 1889) Chitwood, 1949

Syn. *Anguillula arenaria* Neal, 1889
Tylenchus arenarius (Neal, 1889) Cobb, 1890
M. arenaria thamesi (Neal, 1889) Chitwood *et al.*, 1952
M. thamesi (Chitwood *et al.*, 1952) Goodey, 1963

Meloidogyne arenaria

Neal (1889); *Bull. U.S. Bur. Ent.* No. 20: 1-31.
   Chitwood (1949); *Proc. Helminth. Soc. Washington* 16: 90-104.
   Gillard (1961); *Meded. LandbHoogesch., Gent* 26: 515-646.
   Whitehead (1968); *Transactions of the Zoological Society of London* 31: 263-401.
   Orton Williams (1975; *Meloidogyne arenaria. C.I.H. Descriptions.* Set 5, No. 62. CAB, St Albans, UK.
   Cliff & Hirschmann (1985); *Journal of Nematology* 17: 445-459.
   Jepson (1987); *Identification of root-knot nematodes* (Meloidogyne *species*). Wallingford, UK: CAB International. 265 pp.
   Eisenback & Hirschmann (1991); In: *Manual of agricultural Nematology*, Ed. W.R. Nickle. New York: Marcel Dekker. pp. 191-274
   Rammah & Hirschmann (1993); *Journal of Nematology* 25: 103-120 & 121-135.

Measurements & Description

For a complete review on *M. arenaria* morphology and morphometrics, see Orton Williams (1975) and Eisenback & Hirschmann (1991).

Hosts

Almost every plant family is parasitized by *M. arenaria* (Goodey *et al.*, 1965), including economically important food crops and ornamentals. Galls range in size from small to relatively large; sometimes the galls are arranged like a string of pearls.

Distribution

M. arenaria is restricted in the field to the southern countries of Europe, like Portugal, Spain, southern France, Italy, Greece etc. In northern Europe it is very often detected in glasshouses.

Type locality & host

Archer, Florida or Lake City, Florida, USA, described from *Arachis hypogaea* L.

Type material

Two paralectotype slides with males, female heads and perineal pattern from the type locality were available for study (Whitehead, 1968).

Etymology

The species epithet means 'sandy' or 'sand inhabitant'.

Remarks

The esterase A1, 2 & 3 types and the malate dehydrogenase N1 type were described by Esbenshade & Triantaphyllou (1985).

Meloidogyne artiellia **Franklin, 1961**

(Fig. 1, 2 & 3)

Meloidogyne artiellia

Franklin (1961); *Journal of Helminthology, R.T. Leiper Supplement* pp. 85-92.
  Whitehead (1968); *Transactions of the Zoological Society of London* 31: 263-401.
  Di Vito *et al*. (1985); *Nematol. medit.* 13: 207-212.
  Jepson (1987); *Identification of root-knot nematodes* (Meloidogyne *species*). Wallingford, UK: CAB International. 265 pp.
  Di Vito & Creco (1988); *Revue Nématol.* 11: 223-227.
  Eisenback & Hirschmann (1991); In: *Manual of agricultural nematology,* Ed. W.R. Nickle. New York: Marcel Dekker. pp. 191-274.
  Greco *et al*. (1992); *Florida Dept. Agric. & Consumer serv., Div. Plant Ind., Nematode Circular*. No. 201.

Measurements

(in glycerine) see Table I

Female

Body pear shaped with broad, short neck and without posterior protuberance. Stylet relatively short, cone straight, knobs relatively small, ovoid and backwardly sloping. S-E pore located between stylet knobs and metacorpus level. Perineal pattern distinct, rounded with fine striae, lateral area with coarse ridges, dorsal arch angular, lateral field obscure. Anus not always covered with cuticular fold.

Male

Head set off from body. Head cap distinct, labial disc not elevated, lateral lips present. Head region without transverse incisures. Short stylet, knobs backwardly sloping and ovoid shaped. DGO to stylet knobs relatively long. Lateral field with four incisures, outer bands areolated.

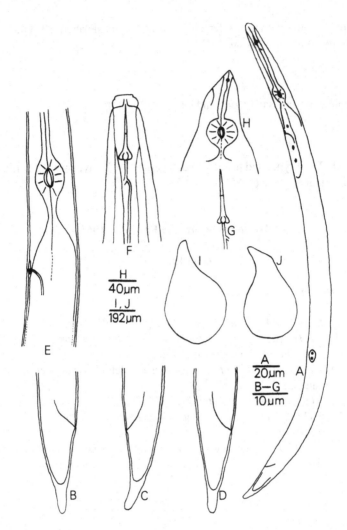

Fig. 1. *Meloidogyne artiellia*. A-E: Second-stage juveniles (lateral). A: Body; B-D: Tails; E: Metacorpus region; F: Male head end (lateral); G-J: Female (lateral). G: Stylet; H: Anterior end; I, J: Body shape (after Franklin, 1961).

Second-stage juvenile

Body relatively short. Hemizonid anterior and adjacent to the S-E pore. Tail relatively short and conical. Rectum rarely inflated. Hyaline tail part very short, ending in rounded tip.

Hosts

In the field detected on *Brassica napus* L., *B. oleracea* L., *Cicer arietinum* L., *Vicia sativa* L., *Avena sativa* L. and *Triticum vulgare* L. Experimental hosts include also other members of the Brassicaceae, Fabaceae and Poaceae (Di Vito *et al.*, 1985). In general *M. artiellia* induces very small galls, sometimes with lateral root proliferation.

Distribution

Reported in Europe from the UK, France, Spain, Italy and Greece (Greco *et al.*, 1992).

Type locality & host

Wells, Norfolk UK, described from *Brassica oleracea* L. var. *capitata.*

Type material

Paratypes deposited at Rothamsted Experimental Station, Harpenden UK, were studied.

Etymology

The meaning of the species is unknown to me.

Remarks

The studied slides are in agreement with the description, except for a shorter observed juvenile and male stylet length, as already noticed by Whitehead (1968) and Jepson (1987).

M. artiellia populations from France, Italy and the UK, all show a malate dehydrogenase N1b type and an unusual esterase M2-VF1 type (two close bands and a weaker one). The malate dehydrogenase type is the same type as was detected for *M. fallax* Karssen, 1996 and for the undescribed *Meloidogyne* sp. 'Rhine', as mentioned in chapter 3. It is interesting that this N1b Mdh type has only been detected in Europe so far.

Based on the present species distribution it is not unlikely that *M. artiellia* will be detected in Belgium, Germany and the southern part of the Netherlands.

As *M. artiellia* second-stage juveniles belong to the smallest *Meloidogyne* juveniles in Europe, they are probably overlooked easily or confused with other small non plant-parasitic nematode genera in soil samples.

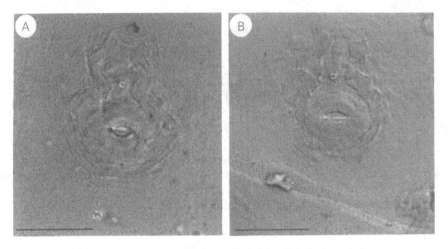

Fig. 2. *Meloidogyne artiellia*. A & B: Female perineal patterns. Bar= 25 μm.

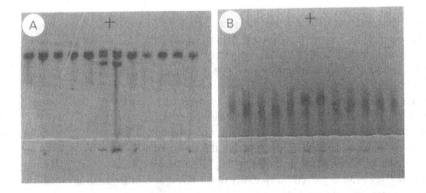

Fig. 3. *Melodogyne artiellia*. Esterase (A) and malate dehydrogenase patterns (B) of populations from Italy (Monopoli), France (Laons) and the UK (lincolnshire); reference: *M. javanica* (China) in the two middle positions.

Table I

Morphometrics of Meloidogyne artiellia *Franklin, 1961*
[*mean ± SD (range); all measurements in µm*]

Character	J2	Males	Females
N	10	6	9
Body length	345±11.6	1008±45	-
	(326-367)	(937-1040)	
Greatest body diameter	14.9±0.4	26.7±1.1	-
	(14.5-15.2)	(25.3-27.8)	
Body diam. at stylet knobs	-	12.0±0.5	-
		(11.4-12.6)	
„ „ „ S-E pore	13.3±0.6	20.1±1.1	-
	(12.6-13.9)	(19.0-22.1)	
„ „ „ anus	8.8±0.6	-	-
	(7.6-9.5)		
Stylet length	12.6±0.3	16.3±0.7	13.9±0.3
	(12.1-13.3)	(15.2-16.4)	(13.3-14.5)
DGO	3.6±0.3	4.9±0.4	4.0±0.3
	(3.2-3.8)	(4.4-5.1)	(3.8-4.4)
S-E pore to anterior end	-	-	31.7±4.4
			(22.1-37.9)
Anterior end to metacorpus	-	56±4.6	54±3.2
		(52-60)	(51-60)
Metacorpus length	-	-	47.3±3.2
			(44.2-53.7)
„ diameter	-	-	45.0±3.5
			(37.9-48.0)
Tail length	22.2±1.8	-	-
	(19.0-25.3)		
Tail terminus length	6.5±1.4	-	-
	(4.4-8.2)		
Anus-primordium	77±6.4	-	-
	(71-92)		
Spicule	-	26.7±0.9	-
		(25.3-27.8)	
Gubernaculum	-	8.2±0.7	-
		(7.6-8.8)	
a	23.2±1.2	37.9±2.9	-
	(21.5-25.3)	(34.5-41.7)	
b"	6.2±0.3	-	-
	(5.8-6.8)		
c	15.6±1.2	-	-
	(13.3-17.2)		
c'	2.6±0.3	-	-
	(2.1-3.1)		
T	-	58±7.3	-
		(46-68)	
(S-E pore/L)x 100	21.1±1.0	11.3±0.6	-
	(19.5-22.6)	(10.8-12.0)	

Meloidogyne chitwoodi Golden, O' Bannon, Santo & Finley, 1980

(Fig. 4, 5 & 6)

Meloidogyne chitwoodi

Golden *et al.* (1980); *Journal of Nematology* 12: 319-327.
O' Bannon *et al.* (1982); *Plant Disease* 66: 1045-1048.
Jepson (1985); *Meloidogyne chitwoodi. C.I.H. Descriptions*. Set 8, No. 106. CAB, St. Albans, UK.
Jepson (1987); *Identification of root-knot nematodes* (Meloidogyne *species*). Wallingford, UK: CAB International. 265 pp.
Eisenback & Hirschmann (1991); In: *Manual of agricultural nematology*, Ed. W.R. Nickle. New York: Marcel Dekker. pp. 191-274.
Pinkerton *et al.* (1991); *Journal of Nematology* 23: 283-290.

Measurements

(in glycerine) see Table II

Female

Body rounded with short neck and slight posterior protuberance. Stylet relatively short, cone slightly curved dorsally, knobs small, oval to irregularly shaped, backwardly sloping. S-E pore located between stylet knobs and metacorpus level. Perineal pattern rounded to oval, striae relatively coarse, dorsal arch ranging from low and rounded to relatively high and angular, lateral lines weakly visible.

Male

Body anteriorly tapering. Head not set off from body. Head cap rounded, labial disc elevated, lateral lips present. Head region without transverse incisures. Stylet relatively short, knobs small, oval to irregularly shaped, backwardly sloping. DGO to stylet knobs relatively short. Lateral field with four incisures, weakly areolated.

Second-stage juvenile

Body moderately long. Hemizonid anterior and adjacent to the S-E pore. Tail conical, of medium size, rectum inflated. Hyaline tail part relatively short, anterior region clearly delimitated, tail tip bluntly rounded.

Hosts

M. chitwoodi is able to parasitize many mono- and dicotyledonous hosts, including economical important food crops as potato, wheat and maize (Jepson, 1987).

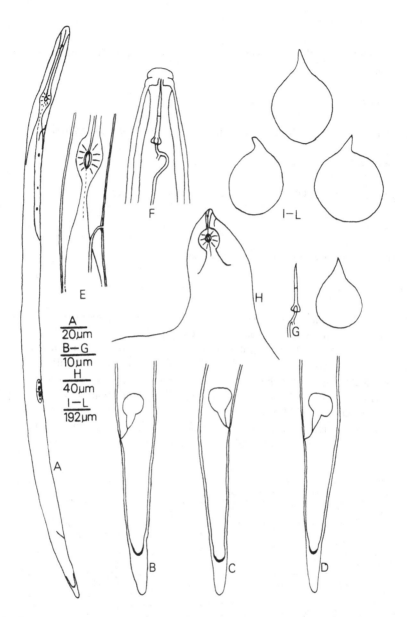

Fig. 4. *Meloidogyne chitwoodi*. A-E: Second-stage juvenile (lateral). A: Body; B-D: Tails; E: Metacorpus region; F: Male head end (lateral); G-L: Female (lateral). G: Stylet; H: Anterior end; I-L: Body shape.

Distribution

In Europe detected in the Netherlands, Belgium, Germany (few locations) and Portugal (one location).

Type locality & host

A field near Quincy, Washington USA, described from *Solanum tuberosum* L.

Type material

Paratypes deposited at USDA, Beltsville, Maryland USA, were studied.

Etymology

This species is named after Dr. B.G. Chitwood.

Remarks

The studied slides are in agreement with the description. The described vesicle-like structures in the anterior part of the female metacorpus were not observed, however in fresly fixed material they are clearly visible (see also *M. fallax*). No morphological and isozyme differences where noticed between *M. chitwoodi* populations from Middle- and North-America (including the three host races) and studied European populations.

M. chitwoodi was added to the European list of quarantine organism in 1998, to prevent further distribution within Europe.

All studied populations reveal an esterase S1 type and a malate dehydrogenase N1a type, these types are in agreement with Esbenshade & Triantaphyllou (1985).

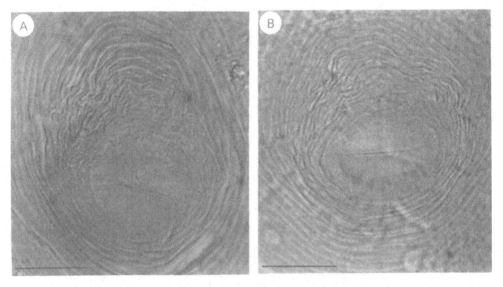

Fig. 5. *Meloidogyne chitwoodi*. A & B: Female perineal patterns. Bar= 25 µm.

Table II

Morphometrics of Meloidogyne chitwoodi *Golden et al., 1980*
[mean ± SD (range); all measurements in μm]

Character	J2	Males	Females
N	10	6	6
Body length	380±11.5	983±60	-
	(362-394)	(896-1070)	
Greatest body diameter	13.1±0.5	29.7±2.6	-
	(12.6-13.9)	(27.8-34.1)	
Body diam. at stylet knobs	-	15.0±1.0	
		(13.9-15.8)	
„ „ „ S-E pore	11.8±0.3	24.5±2.9	
	(11.4-12.0)	(21.5-28.4)	
„ „ „ anus	9.4±0.4	-	-
	(8.9-10.1)		
Stylet length	9.7±0.3	17.7±0.4	11.8±0.3
	(9.5-10.1)	(17.1-18.3)	(11.4-12.0)
DGO	3.4±0.4	3.3±0.5	4.1±0.3
	(2.5-3.8)	(2.5-3.8)	(3.8-4.4)
S-E pore to anterior end	-	-	24.7±3.9
			(19.6-28.4)
Anterior end to metacorpus	-	61±4.3	47±7.4
		(57-68)	(41-58)
Metacorpus length	-	-	41.4±1.8
			(37.9-42.3)
„ diameter	-	-	38.1±4.4
			(29.7-41.1)
Tail length	43.2±1.6	-	
	(39.8-44.8)		
Tail terminus length	10.9±0.8	-	
	(8.9-12.0)		
Anus-primordium	93±8.5	-	-
	(82-109)		
Spicule	-	27.1±0.6	
		(26.5-27.8)	
Gubernaculum	-	8.9±0.7	
		(8.2-9.5)	
a	29.1±1.3	33.2±2.1	
	(26.0-31.0)	(31.4-36.8)	
b"	7.6±0.9	-	
	(5.7-8.8)		
c	8.8±0.4		
	(8.1-9.5)		
c'	4.6±0.2	-	
	(4.2-5.0)		
T	-	57±7.7	
		(48-69)	
(S-E pore/L)x 100	18.5±0.8	11.7±1.4	
	(18.0-19.1)	(10.2-13.6)	

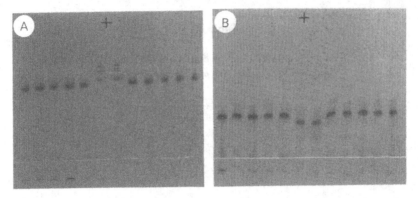

Fig. 6. *Meloidogyne chitwoodi*. Esterase (A) and malate dehydrogenase patterns (B) of populations from the Netherlands (Rips and Smilde) and Belgium (?); reference: *M. javanica* (China) in the two middle positions of the gel.

Meloidogyne fallax Karssen, 1996

(Fig. 7-15)

Syn. *Meloidogyne chitwoodi* B-type (van Meggelen *et al.*,1994)

Meloidogyne fallax

Karssen (1996); *Fundam. appl. Nematol.* 19: 593-599, Erratum (1997); *F.A.N.* 20: 633.
  Karssen (1995); *Nematologica* 41: 314-315.
  Zijlstra *et al.* (1995); *Phytopathology* 85: 1231-1237.
  Brinkman *et al.* (1996); *Anz. Schädlingskde., Planzenschutz, Umweltschutz.* 69: 127-129.
  Petersen & Vrain (1996); *Fundam. appl. Nematol.* 19: 601-605.
  Janssen *et al.* (1996); *Eur. J. Plant Pathol.* 102: 859-865.
  Janssen *et al.* (1996); *Euphytica* 92: 287-294.
  Janssen (1997); *Pesticide Outlook* 8: 29-31.
  Janssen *et al.* (1997); *Fundam. appl. Nematol.* 20: 449-457.
  Janssen *et al.* (1997); *Theoritical appl. Genetics.* 94: 692-700.
  Van der Beek & Karssen (1997); *Phytopathology* 87: 1061-1066.
  Van der Beek *et al.* (1997); *Fundam. appl. Nematol.* 20: 513-520.
  Zijlstra (1997); *Fundam. appl. Nematol.* 20: 505-511.
  Castagnone-Serone (1998); *Mol. Biol. Evol.* 15: 1115-1122.
  Schmitz *et al.* (1988); *Nachrichtenbl. Deut. Pflanzenschutzd.* 50: 310-317.

Preface

In 1992 a field plot experiment was conducted near Baexem, the Netherlands, to assess the host suitability of *Meloidogyne chitwoodi* Golden *et al.*, 1980 on different crops. *Zea mays* L., a good host for *M. chitwoodi* (O'Bannon *et al.*, 1982; Jepson, 1987) was recorded as a non- to poor host. A critical re-examination of second-stage juveniles, compared with *M. chitwoodi* paratypes, indicated differences in body-, tail- and hyaline tail terminus length.

In the autumn of the same year the Baexem population was studied biochemically. A unique malate dehydrogenase (Mdh) pattern was detected (indicated as *M. chitwoodi* B-type), deviating from *M. chitwoodi* phenotypes (Karssen, 1994; van Meggelen *et al.*, 1994). A potato root-knot nematode survey in 1993 revealed also seven other isozyme deviating populations (B-types) in the Baexem region.

After a detailed study, of all known B-type populations, they were considered as morphologically and biologically different from *M. chitwoodi* (Karssen, 1995). The nematode differs also from other known root-knot nematodes. Because of these differences this nematode was designated as a new species and described as *Meloidogyne fallax* Karssen, 1996. *M. fallax* was added to the European list of quarantine organism in 1998, to prevent further distribution within Europe.

Measurements

(in TAF) see Table III

Female

Body annulated, pearly white, globular to pear shaped, with slight posterior protuberance and distinct neck region projecting from the body axis up at an angle of 90° to one side. Head region set off from body, marked with one or two annules. Head cap distinct but variable in shape; labial disc slightly elevated. Cephalic framework weakly sclerotized; vestibule extension distinct.

Stylet cone dorsally curved and shaft cylindrical; knobs large, rounded to transversely ovoid, slightly sloping posteriorly from the shaft. Excretory pore located between head end and metacorpus levels. One or two large vesicle-like structures, and several smaller ones located along the lumen lining. Pharyngeal glands variable in size and shape.

Perineal pattern ovoid to oval shaped, sometimes rectangular; dorsal arch ranging from low to moderately high, with coarse striae. Tail terminus indistinct without punctations. Phasmids small and difficult to observe. Perivulval area devoid of striae. Lateral lines indistinct (LM), appearing as a weak indentation under SEM, increasing towards the tail terminus region and resulting in a relatively large area without striae. Ventral pattern region oval to angular shaped; striae moderately coarse.

Male

Body vermiform, slightly tapering anteriorly, bluntly rounded posteriorly. Cuticle with distinct transverse striae. Lateral field with four incisures; outer ones irregularly aerolated; a fifth broken longitudinal incisure is rarely present near mid-body.

Head slightly set off, with a single post-labial annule (sometimes called head region) usually partly subdivided with a transverse incisure. Labial disc rounded, elevated and fused with medial lips. Prestoma hexagonal in shape with six inner cephalic sensilla adjacent to the rim. Medial lips crescent shaped with raised edges at lateral sides. Four cephalic sensilla small and marked by cuticular depressions on the medial lips. Amphidial openings appear as elongated slits between labial disc and medium sized lateral lips. Cephalic framework moderately sclerotized, vestibule extension distinct.

Stylet cone straight; shaft cylindrical; knobs large and rounded, set off from the shaft. Pharynx with slender procorpus, metacorpus oval shaped with pronounced valve. Ventrally overlapping pharyngeal gland lobe variable in length. Hemizonid, 2-3 μm in length, two to four annules anterior to excretory pore.

Testis usually long, monorchic, with reflexed or outstretched germinal zone. Tail short and twisted. Spicules slender, curved ventrally; gubernaculum slightly crescent shaped. Phasmids located anterior to cloaca.

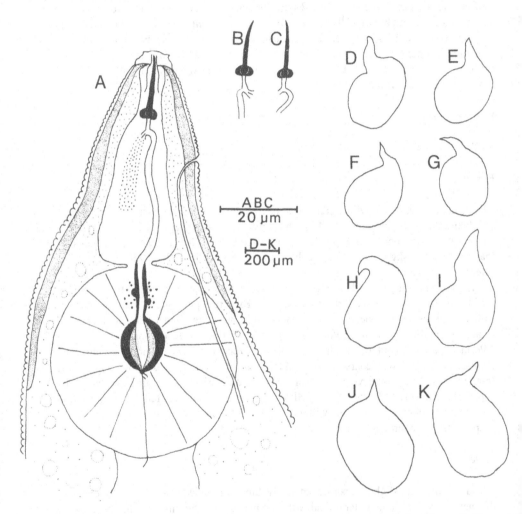

Fig. 7. *Meloidogyne fallax* females. A: Pharyngeal region (lateral); B-C: Stylets (lateral); D-K: Body shapes.

Second-stage juveniles

Body moderately long, vermiform, tapering at both ends but posteriorly more than anteriorly. Body annules small but distinct. Lateral field with four incisures, not areolated. Head region truncate, slightly set off from body. Head cap low and narrower than head region. Cephalic framework weakly sclerotized, vestibule extension distinct.

Stylet slender and moderately long, cone straight; shaft cylindrical; knobs distinct, rounded and set off from the shaft. Pharynx with faintly outlined procorpus and oval shaped metacorpus with distinct valve. Pharyngeal gland lobe variable in length, overlapping intestine ventrally. Hemizonid distinct at the same level with the excretory pore.

Moderately sized tail, gradually tapering until hyaline tail terminus, with inflated proctodeum. Phasmids difficult to observe, small, slightly posterior to anus. A rounded hypodermis marked the anterior position of the smooth hyaline tail terminus ending in a broadly rounded tip. Terminus generally marked by faint cuticular constrictions.

Eggs (n=30)

Length 90-104 µm (94 ± 3.4); width 34.1-44.2 µm (38.9 ± 3.2); length/width ratio 2.1-2.9 (2.4 ± 0.2).

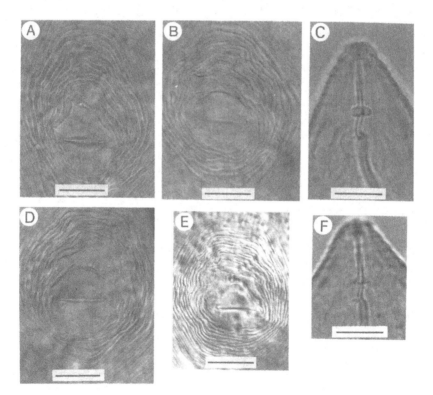

Fig. 8. *Meloidogyne fallax* (A-D) and *M. chitwoodi* (E-F) females. A-B & D-E: Perineal patterns; C & F: Head end (lateral); A-B & D-E: scale bar= 25 µm; C & F: scale bar= 10 µm.

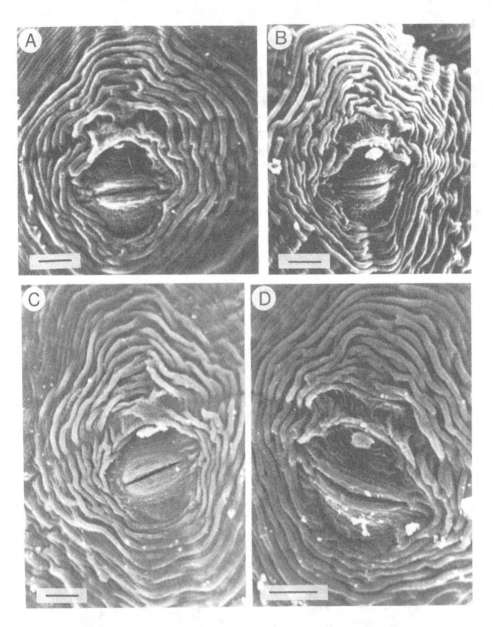

Fig. 9. *Meloidogyne fallax* females. A-D: Perineal patterns; scale bar= 10 μm.

Hosts

In the field detected on different economical important crops like potato and carrot, recent host tests indicate that it is able to parasitize, like *M. chitwoodi*, on a wide range of mono- and dicotyledonous plants. Both root-knot nematodes share hosts, but also species specific host have been found. *M. fallax* induces on most plants relatively small galls.

Host preference studies showed that *Phaseolus vulgaris* L. cv's Iprin, Strike and Groffy, *Zea mays* L., *Potentilla fruticosa* L. and *Erica cinerea* L. were good hosts for *M. chitwoodi* but not for *M. fallax*. On the other hand *M. fallax* reproduced well on *Oenothera erythrosepala* Borb., *Phacelia tanacetifolia* Bentham, *Dicentra spectabilis* (L.) Lem. and *Hemero callis* cv Rajah, while *M. chitwoodi* reproduced not or poorly (Brinkman *et al.*, 1996).

Distribution

Detected so far in the southern part of the Netherlands (Fig. 15), Belgium, Germany (one location) and France (one location).

Type locality and host

Described from roots of tomato (*Lycopersicon esculentum* Mill.). The nematodes were originally isolated from infected roots of black salsify (*Scorzonera hispanica* L. cv. Lange Jan) from arable land one mile north of Baexem, province of Limburg, the Netherlands.

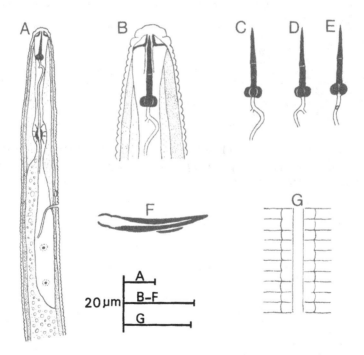

Fig. 10. *Meloidogyne fallax* males (lateral). A: Pharyngeal region; B: Head end; C-E: Stylets (E: ventral); F: Spicule and gubernaculum; G: Lateral field at mid-body.

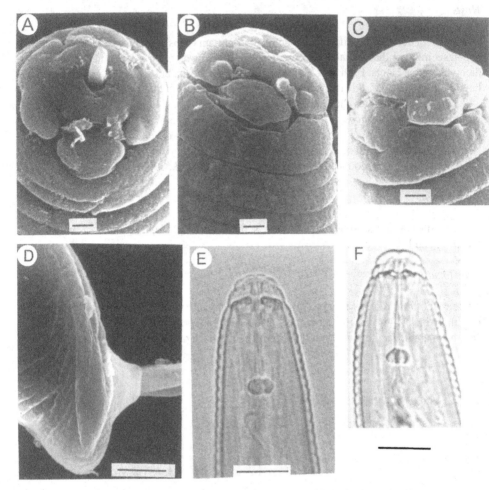

Fig. 11. *Meloidogyne fallax* (A-B & D-E) and *M. chitwoodi* (C & F) males (lateral). A: Cephalic region (face view); B-C: Cephalic region; D: Tail; E & F: Cephalic region; scale bar= 1 µm (A-C), 5 µm (D), 10 µm (E-F).

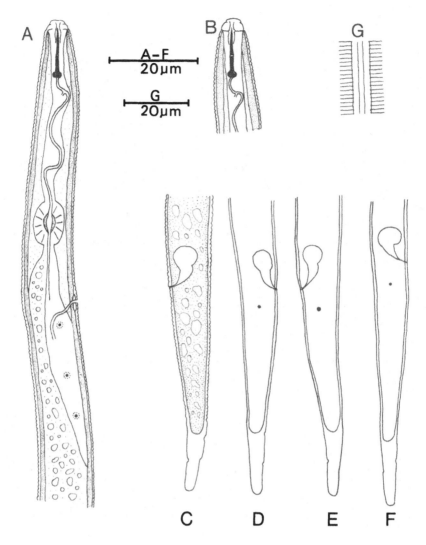

Fig. 12. *Meloidogyne fallax* second-stage juveniles (lateral). A: Pharyngeal region; B: Head end; C-F: Tail shape variation; G: Lateral field at mid-body.

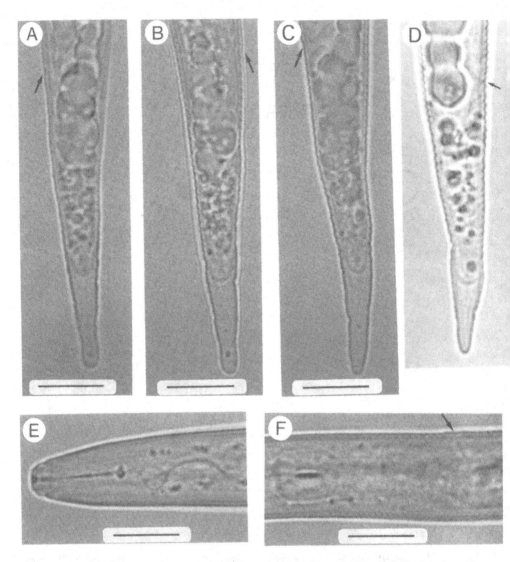

Fig. 13. *Meloidogyne fallax* (A-C, E-F) and *M. chitwoodi* (D) second-stage juveniles (lateral). A-D: Tail variation (arrow= anus); E: Head end; F: Meta- and postcorpus region (arrow= hemizonid); scale bar= 10 µm.

Type material

Holotype. Female on slide WT 3127, collection of Agricultural University, Wageningen the Netherlands. *Paratypes*. Two female perineal patterns and heads, two males and five J2's deposited at each of the following nematode collections: Agricultural University, Wageningen the Netherlands (WT 3128-313O); Instituut voor Dierkunde, Rijksuniversiteit, Gent Belgium; Rothamsted Experimental Station, Harpenden UK.

Etymology

The species epithet refers to the misleading morphological resemblance to *M. chitwoodi*.

Remarks

M. fallax is characterized by a dorsally curved female stylet of 14.5 μm (13.9-15.2) with rounded set off stylet knobs. Oval shaped perineal pattern with coarse striae and moderately high dorsal arch. Male stylet length 19.6 μm (18.9-20.9) with prominent set off rounded knobs. The labial disc is elevated, crescent shaped medial lips raised at lateral side and distinct lateral lips. The J2's hemizonid is at the same level with the excretory pore. Tail and hyaline tail length 49.3 μm (46.1-55.6) and 13.5 μm (12.1-15.8) respectively.

This species differs from the morphologically close related *M. chitwoodi* Golden *et al.*, 1980 by greater female and male stylet length, absence of small, irregular outlined male and female stylet knobs (Eisenback & Hirschmann, 1991), male labial disk elevated, longer juvenile body-, tail- and hyaline tail length, different hyaline tail shape, hemizonid position, esterase and malate dehydrogenase patterns (Fig. 8, 11 and 13); from *M. hapla* Chitwood, 1949 by the absence of fine, smooth striae, rounded and flattened dorsal arch and tail area punctations in the female perineal pattern, broader J2 tail and tail terminus with distinct hyaline part, shorter female and male stylet length, and the absence of small rounded stylet knobs; from *M. artiellia* Franklin, 1961 and *M. ardenensis* Santos, 1968 by much greater J2 body-, tail- and hyaline tail length and by hemizonid position relative to excretory pore; from *M. naasi* Franklin, 1965 by smaller J2 body-, tail- and hyaline tail length and the absence of vesicles in the juvenile metacorpus.

M. fallax reproduces by facultative meiotic parthenogenesis, the haploid chromosome number is n=18 (Dr. H. v.d. Beek, pers. comm.). It is also characterized by a unique malate dehydrogenase (Mdh) pattern, not described by Esbenshade and Triantaphyllou (1987), and the lack of any major esterase (Est) band (Fig. 14). All known *M. fallax* populations share this rare malate dehydrogenase N1b type and 'null' esterase type; prolonged esterase staining, of 1½ hour, revealed a very weak three-banded pattern named F3 (van der Beek & Karssen, 1997). In combination they are useful to differentiate it from other known *Meloidogyne* species. Beside the mentioned difference in isozyme patterns between *M. chitwoodi* and *M. fallax*, biochemical differentiation was also confirmed by restriction analysis of ribosomal (ITS) DNA (Zijlstra *et al.*, 1995; Petersen & Vrain, 1996).

Table III

Morphometrics of Meloidogyne fallax *Karssen, 1996*
[mean ± SD (range); n= 30; all measurements in µm]

character	Females	Males	J2
L	49+74.9	1171+193.6	403+15.2
	(404-720)	(736-1520)	(381-435)
Greatest body diam.	362+57.7	30.6+2.1	14.3+0.7
	(256-464)	(27.2-43.8)	(13.3-16.4)
Body diam. at stylet knobs	-	17.7+0.6	-
		(16.4-19.0)	
Body diam. at excr. pore	-	26.2+1.7	-
		(23.4-29.7)	
Body diam. at anus	-	-	10.4 + 0.4
			(9.5-10.7)
Head region height	-	4.6+0.3	2.7+0.4
		(4.4-5.1)	(1.9-3.2)
Head region diam.	-	10.7+0.7	5.5+0.3
		(9.5-12.0)	(5.1-6.3)
Neck length	150+33.0	-	-
	(96-225)		
Neck diam.	98+23.6	-	-
	(65-160)		
Stylet	14.5+0.4	19.6+0.8	10.8+0.4
	(13.9-15.2)	(18.9-20.9)	(10.1-11.4)
Stylet base-ant. end	-	-	14.6+0.7
			(13.9-16.4)
Stylet cone	-	10.1+0.5	-
		(9.5-12.0)	
Stylet shaft and knobs	-	8.9+0.5	5.5+0.4
		(8.2-9.5)	(5.1-6.3)
Stylet knob height	2.3+0.3	3.0+0.3	1.5+0.3
	(2.0-2.5)	(2.5-3.2)	(1.3-1.9)
Stylet knob width	4.2+0.3	4.9+0.4	2.3+0.3
	(3.8-4.4)	(3.8-5.1)	(1.9-2.5)
DGO	4.3+0.5	4.4+0.7	3.5+0.3
	(3.8-6.3)	(3.2-5.7)	(3.2-3.8)
Ant. end to metacorpus	-	65+4.1	48+3.5
		(59-73)	(44-54)
Metacorpus length	41.9+3.3	-	-
	(34.8-47.4)		
Metacorpus diam.	39.6+3.8	-	-
	(31.6-44.9)		
Metacorpus valve length	12.0+0.9	-	4.0+0.3
	(10.1-13.9)		(3.2-3.8)
Metacorpus valve width	9.7+0.6	-	3.3+0.2
	(8.2-11.4)		(3.2-3.8)
Excretory pore-ant. end	22.5+5.3	121+11.4	69+3.4
	(12.6-32.9)	(95-140)	(63-77)
Tail	-	9.2+1.4	49.3+2.2
		(7.6-12.1)	(46.1-55.6)
Tail terminus length	-	-	13.5+1.0
			(12.2-15.8)

Phasmids-post. end	-	11.7±1.5 (9.5-15.2)	-
Spicule	-	26.6±2.0 (22.1-29.7)	-
Gubernaculum	-	7.7±0.5 (7.0-8.5)	-
Testis	-	497±144.1 (316-695)	-
Vulva slit length	24.7±1.8 (20.2-28.4)	-	-
Vulva-anus distance	15.9±1.8 (12.6-19.0)	-	-
a	1.4±0.3 (0.9-2.0)	38.2±6.8 (21.2-53.5)	28.1±1.7 (23.8-40.4)
c	-	128±28.5 (83-202)	8.2±0.5 (6.9-8.6)
c'	-	-	4.8±0.3 (4.3-5.3)
T	-	42.4±8.4 (24.4-62.1)	-
Body length/neck length	3.3±0.9 (1.9-5.6)	-	-
Body length/ant. end to metacorpus valve	-	-	8.4±0.7 (6.4-9.3)
Stylet knob width/height	1.9±0.2 (1.5-2.2)	1.6±0.2 (1.4-2.0)	-
Metacorpus length/width	1.1±0.1 (0.9-1.2)	-	-
Excretory pore/L x 100	-	10.3±1.9 (8.3-12.9)	17.2±1.1 (16.1-19.3)

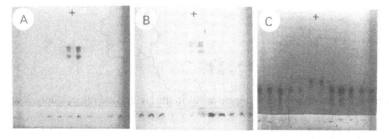

Fig. 14. *Meloidogyne fallax*. Esterase (A), after prolonged staining (B) and malate dehydrogenase patterns (C) of the type population from the Netherlands (Baexem); reference: *M. javanica* (China) in the two middle positions of the gel.

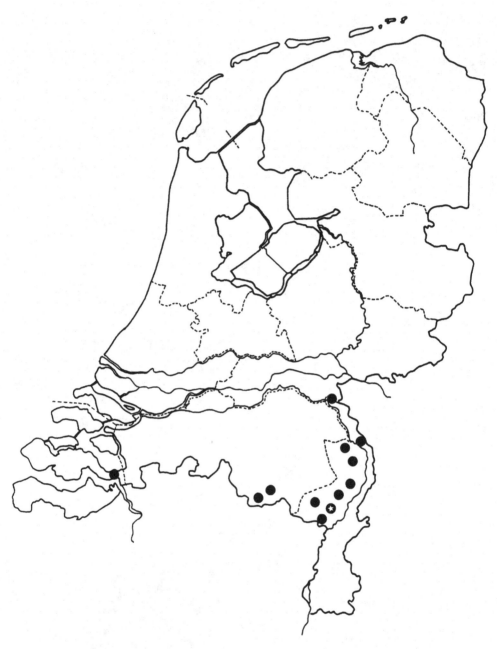

Fig. 15. Distribution pattern of *Meloidogyne fallax* in the Netherlands (black dots; black dot with star is type locality).

Meloidogyne hispanica Hirschmann, 1986

(Fig. 16 & 17)

Meloidogyne hispanica

Hirschmann (1986); *Journal of Nematology* 18: 520-532.
  Dalmasso & Bergé (1978); *Journal of Nematology* 10: 323-332.

Measurements

(in glycerine) see Table IV

Female

Body rounded with short neck and without a posterior protuberance. Stylet relatively short, cone slightly curved dorsally, knobs robust, rounded and set off. S-E pore located between stylet knobs and metacorpus level. Perineal pattern oval with coarse striae, dorsal arch ranging from low to relatively high and angular, lateral lines weakly visible.

Male

Head high and set off from body. Head cap rounded, labial disc elevated, lateral lips present. Head region with transverse incisures. Stylet relatively long, knobs robust, rounded and set off. DGO to stylet knobs relatively short. Lateral field with four incisures, areolated.

Second-stage juvenile

Body of medium length. Hemizonid ranging from anterior to anterior-adjacent to the S-E pore. Tail slightly curved ventrally, slender, of medium length, rectum inflated. Hyaline tail part not clearly delimitated anteriorly, tail tip finely rounded.

Hosts

Type population found on *Prunus persica* Batsch. In greenhouse tests, it reproduced on tomato, while tobacco, pepper and watermelon were slightly infected (Hirschmann, 1986). In 1995, the Plant Protection Service (the Netherlands) received an unknown root-knot nematode from Spain and identified it as *M. hispanica*. This population was detected on *Beta vulgaris* L.; in the greenhouse it reproduced on *Lycopersicon esculentum* Mill. and *Triticum aestivum* L.

Distribution

In Europe known from a few local infections in Spain and Portugal (Hirschmann, 1986).

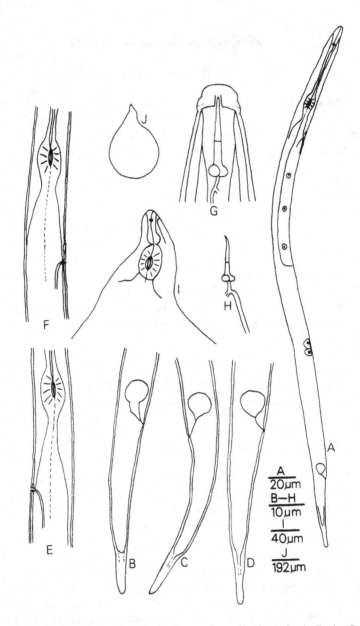

Fig. 16. *Meloidogyne hispanica*. A-F: Second-stage juvenile (lateral). A: Body; B-D: Tails; E, F: Metacorpus region; G: Male head end (lateral); H-J: Female (lateral). H: Stylet; I: Anterior end; J: Body shape.

Type locality & host

Peach orchard, Seville district (Andalusia) Spain. Although the description mentioned *Prunus persica* as the type host, *M. hispanica* was maintained and described from *Lycopersicon esculentum*, the correct type host.

Type material

Paratypes deposited at USDA, Beltsville, Maryland USA, were studied.

Etymology

The species name refers to Spain.

Remarks

The studied slides are in agreement with the description. The small rounded structures, as originally illustrated between the female DGO and the stylet knobs, where not observed.

I have placed this species in the group of root-knot nematodes parasitzing mono- and dicotyledonous field hosts, despite the fact that it has never been detected in the field on a monocotyledonous host. Very little is known about hosts and distribution of *M. hispanica* in Europe. Based on enzymes and morphology it is very close related to *M. incognita* (Dalmasso & Bergé, 1979).

Esbenshade & Triantaphyllou (1985) described an esterase S2-M1 type and a malate dehydrogenase N1 type for *M. hispanica*.

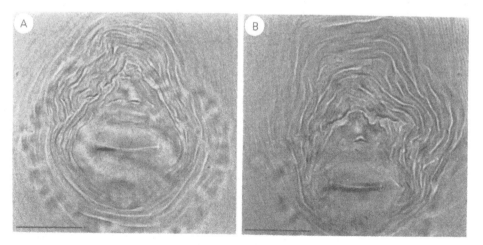

Fig. 17. *Meloidogyne hispanica*. A & B: Female perineal patterns. Bar= 25 μm.

Table IV

Morphometrics of Meloidogyne hispanica *Hirschmann, 1986*
[mean ± SD (range); all measurements in µm]

Character	J2	Males	Females
N	10	4	1
Body length	397±12.3	1928±75	493
	(378-416)	(1856-2000)	
Greatest body diameter	12.5±0.4	34.9±1.5	346
	(12.0-13.3)	(32.9-36.0)	
Body diam. at stylet knobs	-	19.0±0.8	-
		(18.3-19.6)	
„ „ „ S-E pore	11.8±0.5	30.4±2.1	-
	(11.4-12.6)	(27.2-31.6)	
„ „ „ anus	9.3±0.8	-	-
	(8.2-10.7)		
Stylet length	10.2±0.3	22.6±0.6	13.9
	(10.1-10.7)	(22.1-23.4)	
DGO	3.0±0.3	3.0±0.4	3.2
	(2.5-3.2)	(2.5-3.2)	
S-E pore to anterior end	-	-	13.3
Anterior end to metacorpus	-	95±6.0	51
		(89-101)	
Metacorpus length	-	-	41.1
„ diameter	-	-	32.2
Tail length	44.4±2.9	-	-
	(38.6-50.6)		
Tail terminus length	12.6±0.7	-	-
	(11.4-13.3)		
Anus-primordium	91±4.1	-	-
	(85-95)		
Spicule	-	32.6±0.8	-
		(31.6-33.5)	
Gubernaculum	-	8.1±0.3	-
		(7.6-8.2)	
a	31.8±1.2	55±3.7	-
	(30.2-34.7)	(52-60)	
b"	7.3±0.2	-	
	(6.9-7.6)		
c	8.9±0.6	-	-
	(8.5-10.1)		
c'	4.8±0.6	-	-
	(4.1-6.2)		
T	-	34.7±7.3	-
		(27.6-41.9)	
(S-E pore/L)x 100	21.0±1.7	8.7±0.4	-
	(19.9-25.5)	(8.2-9.2)	

Meloidogyne incognita (Kofoid & White, 1919) Chitwood, 1949

Syn. *Oxyuris incognita* Kofoid & White, 1919
Heterodera incognita (Kofoid & White, 1919) Sandground, 1923
M. incognita incognita (Kofoid & White, 1919) Chitwood, 1949
M. incognita acrita Chitwood, 1949
M. incognita inornata Lordello, 1956
M. acrita (Chitwood, 1949) Esser *et al.*, 1976
M. elegans da Ponte, 1977
M. grahami Golden & Slana, 1978
M. incognita wartellei Golden & Birchfield, 1978
M. incognita grahami (Golden & Slana, 1978) Jepson, 1987

Meloidogyne incognita

Kofoid & White (1919); *Journal of the American Medical Association* 72: 567-569.
Chitwood (1949); *Proc. Helminth. Soc. Washington* 16: 90-104.
Triantaphyllou & Sasser (1960); *Phytopathology* 50: 724-735.
Gillard (1961); *Meded. LandbHoogesch., Gent* 26: 515-646.
Whitehead (1968); *Transactions of the Zoological Society of London* 31: 263-401.
Orton Williams (1973); *Meloidogyne incognita. C.I.H. Descriptions.* Set 2, No. 18. CAB, St. Albans, UK.
Jepson (1987); *Identification of root-knot nematodes* (Meloidogyne *species*). Wallingford, UK: CAB International. 265 pp.
Eisenback & Hirschmann (1991); In: *Manual of agricultural Nematology*, Ed. W.R. Nickle. New York: Marcel Dekker. pp. 191-274.

Measurements & Description

See Orton Williams (1973) and Eisenback & Hirschmann (1991) for a complete review of *M. incognita* morphology and morphometrics.

Hosts

Goodey *et al.* (1965) mentioned about 700 plant hosts and varieties parasitized by *M. incognita*, including most economic important crops. Galls are ranging in size from small to very large.

Distribution

In the field restricted to the southern part of Europe, and detected very often in greenhouses in northern Europe.

Type locality & host

Field near El Paso, Texas USA, described from *Daucus carota* L.

Type material

Only one paralectotype, with perineal patterns from the type locality, was available for study (Whitehead, 1968).

Etymology

The species name means 'the unknown', referring to the fact that it was originally described as *Oxyuris* eggs from the faeces of soldiers from Texas USA (see chapter 2).

Remarks

The esterase I1 type and the malate dehydrogenase N1 type were described by Esbenshade & Triantaphyllou (1985).

Meloidogyne javanica (Treub, 1885) Chitwood, 1949

Syn. *Heterodera javanica* Treub, 1885
Tylenchus (Heterodera) javanica (Treub, 1885) Cobb, 1890
Anguillula javanica (Treub, 1885) Lavergne, 1901
M. javanica javanica (Treub, 1885) Chitwood, 1949
M. javanica bauruensis Lordello, 1956
M. lucknowica Singh, 1969
M. lordelloi da Ponte, 1969
M. bauruensis (Lordello, 1956) Esser *et al.*, 1976

Meloidogyne javanica

Treub (1885); *Mededeelingen 's lands plantenuin* 2: 1-39.
 Chitwood (1949); *Proc. Helminth. Soc. Washington* 16: 90-104.
 Gillard (1961); *Meded. LandbHoogesch., Gent* 26: 515-646.
 Whitehead (1968); *Transactions of the Zoological Society of London* 31: 263-401.
 Orton Williams (1972); *Meloidogyne javanica. C.I.H. Descriptions.* Set 1, No. 3. CAB, St. Albans, UK.
 Jepson (1987); *Identification of root-knot nematodes* (Meloidogyne *species*). Wallingford, UK: CAB International. 265 pp.
 Eisenback & Hirschmann (1991); In: *Manual of agricultural Nematology,* Ed. W.R. Nickle. New York: Marcel Dekker. pp. 191-274.

Measurements & Description

See Orton Williams (1972) and Eisenback & Hirschmann (1991) for a complete review of *M. javanica* morphology and morphometrics.

Hosts

A wide range of plant families is parasitized by *M. javanica* (Goodey *et al.*, 1965), including economic important crops. Galls ranging in size from small to relatively large.

Distribution

In the field restricted to the southern part of Europe. In northern Europe, very often detected in greenhouses.

Type locality & host

Buitenzorg, Java Indonesia, described from *Saccharum officinarum* L.

Type material

No original type material was available.

Etymology

The species name refers to Java (Indonesia).

Remarks

The esterase J3 type and the malate dehydrogenase N1 type were described by Esbenshade & Triantaphyllou (1985).

Meloidogyne kirjanovae Terenteva, 1965

(Fig. 18 & 19)

Meloidogyne kirjanovae

Terenteva (1965); *Materialy Nauchnoi Konferentsii Vsesoyuznoi Obshchei. Gelmintologov* 4: 277-281.
   Kirjanova & Krall (1980); *Plant-parasitic nematodes and their control.* Vol. 2. Amerind Publ. Co. Pvt. Ltd., New Delhi. 524 pp.

Measurements

(in glycerine) see Table V

Female

Body rounded with short neck and without a posterior protuberance. Medium-sized stylet, cone curved dorsally, knobs robust, transversely ovoid and set off. S-E pore between stylet knobs and metacorpus level. Perineal pattern oval with relatively high and angular dorsal arch, lateral lines not observed, except for one pattern (Fig. 19B).

Male

Head relatively high, not set off from body. Head cap with elevated labial disc, centrally concave. Head region with transverse incisures. Stylet robust and relatively long, knobs transversely ovoid and set off. DGO to stylet knobs relatively short. Lateral field with four incisures, areolated.

Second-stage juvenile

(From the description). Body moderately long, 392 μm (359-433). DGO to stylet knobs relatively short (3 μm). Tail moderately long, 53 μm (47-60), hyaline tail part relatively short, 8.9 μm (7.5-12).

Hosts

Only known from the type host: *Lycopersicon esculentum* Mill.

Distribution

Known only from the type locality, a greenhouse complex near St. Petersburg Russia.

Type material

All the types deposited in the Zoological Museum, St. Petersburg Russia and BBA, Münster Germany, were studied.

Etymology

The species is named after Dr. E.S. Kirjanova.

Remarks

Kirjanova & Krall (1980) mentioned already that *M. kirjanovae* is possibly the same species as *M. incognita* (Kofoid & White, 1919) Chitwood, 1949.

The male DGO length for *M. kirjanovae* was described as 5.4 μm (4.5-6.9) long by Jepson (1987), in the male identification key on page 233. This is not correct and probably a translation error or mistaken with the male head height. The male DGO length was, in de original Russian publication, described as 2.4 μm (2.3-2.8) long and the male head height as 5.4 μm (4.5-6.9) long.

Based on the studied slides, I consider both as the same species and synonymize *M. kirjanovae* with *M. incognita*.

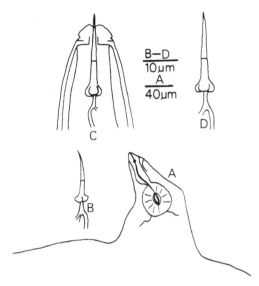

Fig. 18. *Meloidogyne kirjanovae*. A, B: Female (lateral). A: Anterior end; B: Stylet; C, D: Male (lateral). C: Head end; D: Stylet.

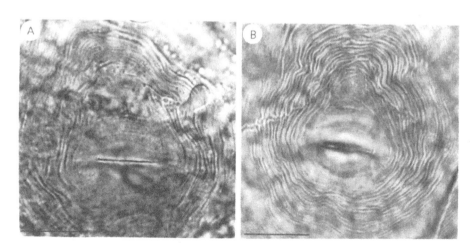

Fig. 19. *Meloidogyne kirjanovae*. A & B: Female perineal patterns. Bar= 25 μm.

Table V

Morphometrics of Meloidogyne kirjanovae *Terenteva, 1965*
[mean ± SD (range); all measurements in µm]

Character	J2	Males	Females
N	-	7	2
Body length	-	1243±335 (885-1548)	
Greatest body diameter	-	42.3±2.3 (40.4-44.9)	
Body diam. at stylet knobs	-	20.3±1.7 (19.0-22.8)	-
„ „ „ S-E pore	-	29.4±0.6 (29.1-30.3)	-
„ „ „ anus	-	-	-
Stylet length	-	22.8±0.6 (22.1-23.4)	14.8±0.4 (14.5-15.1)
DGO	-	2.7±0.4 (2.5-3.2)	2.9±0.5 (2.5-3.2)
S-E pore to anterior end	-	-	22.5±0.5 (22.1-22.8)
Anterior end to metacorpus	-	81±1.8 (79-82)	59±6.7 (54-63)
Metacorpus length	-	-	41.4±1.3 (40.4-42.3)
„ diameter	-	-	40.8±0.5 (40.4-41.1)
Tail length	-	-	-
Tail terminus length	-	-	-
Anus-primordium	-	-	-
Spicule	-	35.6±1.1 (34.8-37.9)	-
Gubernaculum	-	8.5±0.7 (7.6-9.5)	-
a	-	29.2±6.5 (21.9-34.5)	-
T	-	51.3±10.2 (44.1-58.5)	-
(S-E pore/L)x 100	-	11.1±3.3 (8.2-14.7)	-

Discussion

M. chitwoodi & M. fallax

The presence of vesicle-like structures in the metacorpus of *Meloidogyne* females was reported as 'unique' with the description of *M. chitwoodi* Golden *et al.*, 1980, although first reported in *M. kikuyensis* De Grisse, 1961 and *M. oryzae* Maas *et al.*, 1978.

These structures were also recently described in *M. hispanica* Hirschmann, 1986, *M. maritima* Jepson, 1987, *M. konaensis* Eisenback *et al.*, 1995 and *M. fallax* Karssen, 1996. I have also observed these vesicle-like structures in several *M. hapla* populations from Europe. Therefore these vesicle-like structures are not useful as a discriminating character for *M. chitwoodi* identification.

Although *M. fallax* and *M. chitwoodi* are morphologically close related, hybridization tests (van der Beek & Karssen, 1997) and molecular comparison (Petersen & Vrain, 1996; van der Beek *et al.*, 1997) support the species status of *M. fallax*. It is not unlikly that *M. fallax* is a relatively young species, recently split off from *M. chitwoodi* or a common (unknown) ancestor. Both species also occur in the same geographical European region, but are rarely found together in the same field or on the same host. Compared to the distribution of *M. chitwoodi* in Europe, *M. fallax* is limited to a small region in the southern part of the Netherlands and northern Belgium.

Recently I have isolated from samples originating from the type locality 50 distinctly longer *M. fallax* second-stage juveniles. Compared to the described juveniles, they have a longer body, tail and hyaline tail, out of the usual *M. fallax* range. However no morphological differences were observed, suggesting direct polyploidization and a correlation with the cytological polyploid *M. hapla* B-race (chapter 3 & Triantaphyllou, 1984). Direct polyploidization of meiotic and even mitotic parthenogenetic *Meloidogyne* forms is probably rare under field conditions in Europe, only a few *M. hapla* B-races were recorded so far (chapter 3).

Chitwood (1949)

Chitwood (1949) reinstated the genus *Meloidogyne*, described *M. hapla* and redescribed *M. arenaria* (Neal,1889), *M. exigua* (Göldi, 1892), *M. incognita* (Kofoid & White, 1919) and *M. javanica* (Treub, 1885).

Although Chitwood deposited type material (fixed infected roots), he did not designate type specimens. Therefore he mentioned that '*A committee of nematologist in N. America chose type specimens of M. javanica, M. arenaria, M. incognita and M. hapla. .... Lordello will try to obtain type material of M. exigua*' (Chitwood, 1961). As far as I know this has never happened.

Gillard (1961) criticized in detail the nomenclatorial action of Chitwood (1949), mainly because of the way he adopted the early specific names. The original descriptions of *Anguillula arenaria* Neal, 1889, *M. exigua* Göldi, 1892, *Oxyuris incognita* Kofoid & White, 1919 and *Heterodera javanica* Treub, 1885 are so incomplete that these species are not identifiable below generic level, also no type material or specimens were left.

Whitehead (1968) also criticized this action, based upon the type material of Chitwood (1949) and additional collected material he redescribed the Chitwood species and designated lecto- and paralectotypes. These types were deposited in 1968 in the Rothamsted Nematology Collection, Harpenden UK, one year later the lectotypes and some paralectotypes were sent for deposit to the USDA Nematology Collection in Beltsville USA, were they are still kept (Dr. A.M. Golden, pers. comm.).

Beside the redescribed *Meloidogyne* species, Chitwood (1949) described *M. hapla*. About this species he wrote '*The species M. hapla may be a synonym of Anguillula marioni Cornu, 1879. ..... but we are not justified in drawing such a conclusion without material from the type host and locality in France.*' He also postulated the possibility that *M. hapla* was introduced from Europe. *M. marioni* (Cornu, 1879) was described from the roots of *Onobrychis sativa* Lamk. found near Chateauneuf-sur-Loire in France. Although the description is not sufficient for identification on species level, some of the illustrated root-knots strongly resemble typical *M. hapla* galls, i.e. small galls with numerous emerging secondary roots. Nowadays we know that *M. hapla* is widely distributed in France, and it is the predominant *Meloidogyne* species in the Loire valley (H. Marzin, pers. comm.). However I am not able to really proof that *M. hapla* is *M. marioni*, therefore we must accept the present *species inquirenda* status.

Although the early critism on the nomenclatural action of Chitwood (1949) is correct, in the interest of stability it is not wise to change the names of *M. arenaria, M. exigua, M. hapla, M. incognita* and *M. javanica* fifty years after their redescription.

I believe that Chitwood, as the founder of *Meloidogyne* taxonomy, choose the best taxonomical approach available to him at that time. He must have realized this, as he wrote the profetic words '*...it seems better to establish the species with new descriptions than to continue the present confusion. It is hoped that the present separations may serve as a basis on which to build more detailed morphological work in the future.*'

Mode of reproduction, polyploidy level, distribution and parasitic behaviour

Within the European *Meloidogyne* species parasitizing mono- and dicotyledons, there are two distinct complex groups. Group I: *M. arenaria, M. hispanica, M. incognita* and *M. javanica* are mitotic parthenogenetic, polyploid (2n > 30 & < 56), world-wide distributed and have a very broad host range (exception, *M. hispanica*?). Group II: *M. chitwoodi* and *M. fallax* are meiotic parthenogenetic, diploid (n= 18), limited in their distribution and host range (Triantaphyllou, 1985). In Europe, group I members are in the field restricted to the southern part, while group II members occur in the more cooler middle and nothern parts.

Although the chromosome number and mode of reproduction is unknown for *M. artiellia*, I hypothesize it also belongs to group II. *M. artiellia* has been reported from both cooler and warmer parts of Europe. However, in the Mediterranean climate, *M. artiellia* is only active during the winter and spring and becomes quiescent during late spring and summer (Greco *et al.*, 1992).

The two groups are morphologically indistinguishable, only the mean juvenile body- and tail length is slightly different (I: 449 & 52 µm; II: 376 & 38 µm) and probably caused by the reported polyploidization effect (chapter 3 & 5).

References

BRINKMAN, H., GOOSENS, J.J.M. & VAN RIEL, H. (1996). Comparitive host suitability of selected crop plants to *Meloidogyne fallax* Karssen, 1996. *Anzeiger für Schädlingskunde, Pflanzenschutz, Umweltschutz* 69, 127-129.

CHITWOOD, B.G. (1949). 'Root-knot nematodes'. Part 1. A revision of the genus *Meloidogyne* Goeldi, 1887. *Proc. Helminth. Soc. Wash.* 16, 90-104.

CHITWOOD, B.G. (1961). Type specimens of pyroid nemic taxa with comments on variability and host range. *Nematologica* 7, 11-12.

CORNU, M. (1879). Etudes sur le *Phylloxera vastatrix*. *Mém. Divers Sav. Acad. Sc. Institut France* 26, 163-175, 328, 339-341.

DI VITO, M., GRECO, N. & ZACCHEO, G. (1985). On the host range of *Meloidogyne artiellia*. *Nematol. Mediterranea* 13, 207-212.

EISENBACK, J.D. & HIRSCHMANN, H. (1991). Root-knot nematodes: *Meloidogyne* species and races. In: *Manual of agricultural nematology*. pp. 191-274. Ed. W.R. Nickle. New York, Marcel Dekker.

ESBENSHADE, P.R. & TRIANTAPHYLLOU, A.C. (1985). Use of enzyme phenotypes for identification of *Meloidogyne* species. *J. Nematol.* 17, 6-20.

ESBENSHADE, P.R. & TRIANTAPHYLLOU, A.C. (1987). Enzymetic relationships and evolution in the genus *Meloidogyne* (Nematoda: Tylenchida). *J. Nematol.* 19, 8-18.

FRANKLIN, M. (1957). Review of the genus *Meloidogyne*. *Nematologica* 2 (Suppl.), 387-397.

GILLARD, A. (1961). Onderzoekingen omtrent de biologie, de verspreiding en de bestrijding van wortelknobbelaaltjes (*Meloidogyne* spp). *Meded. LandbHoogesch., Gent* 26, 515-646.

GOLDEN, A.M., O'BANNON, J.H., SANTO, G.S. & FINLEY, A.M. (1980). Description and SEM observations of *Meloidogyne chitwoodi* n. sp. (Meloidogynidae), a root-knot nematode on potato in the Pacific Northwest. *J. Nematol.* 12, 319-327.

GÖLDI, E.A. (1892). Relatoria sôbre a molestia do cafeiro na provincia da Rio de Janeiro. *Archos Mus. nac., Rio de Janeiro* 8, 7-112.

GOODEY, J.B., FRANKLIN, M.T. & HOOPER, D.J. (1965). *The nematode parasites of plants catalogued under their hosts*. 3rd edn. Farnham House, Farnham Royal, England, C.A.B. 214 pp.

GRECO, N., VOVLAS, N., DI VITO, M. & INSERRA, R.N. (1992). Meloidogyne artiellia: *A root-knot nematode parasite of cereals and other field crops*. Nematology Circular No. 201, Fla. Dept. Agric. & Consumer Serv., Division of Plant Industry.

HIRSCHMANN, H. (1986). *Meloidogyne hispanica* n. sp. (Nematoda: Meloidogynidae), the 'Seville root-knot nematode'. *J. Nematol.* 18, 520-532.

JEPSON, S.B. (1987). *Identification of root-knot nematodes* (Meloidogyne *species*). Wallingford, UK, C.A.B. International. 265 pp.

KARSSEN, G. (1994). The use of isozyme phenotypes for the identification of root-knot nematodes (*Meloidogyne* species). *Versl. en Meded. Pl. ziektek.Dienst Wageningen (Ann. Rep. Diagnostic Center).* 170, 85-88.

KARSSEN, G. (1995). Morphological and biochemical differentiation in *Meloidogyne chitwoodi* populations in The Netherlands. *Nematologica* 41, 314-315 [Abstr.].

KARSSEN, G., VAN HOENSELAAR, T., VERKERK, B. & JANSSEN, R. (1995). Species identification of cyst and root-knot nematodes from potato by electrophoresis of individual females. *Electrophoresis* 16, 105-109.

KARSSEN, G. (1996). Description of *Meloidogyne fallax* n. sp. (Nematoda: Heteroderidae), a root-knot Nematode from The Netherlands. *Fundam. Appl. Nematol.* 19, 593-599.

KIRJANOVA, E.S. & KRALL, E.L. (1980). *Plant-parasitic nematodes and their control.* Vol. II. Amerind Publ. Co. Pvt. Ltd., New Delhi. 524 pp.

O' BANNON, J.H., SANTO, G.S., NYCZEPIR, A.P. (1982). Host range of the Columbia root-knot nematode. *Pl. Dis.* 66, 1045-1048.

ORTON WILLIAMS, K.J. (1972). *Meloidogyne javanica. C.I.H. Descriptions of plant-parasitic nematodes.* Set 1, No. 3, St. Albans, UK, C.A.B.

ORTON WILLIAMS, K.J. (1973). *Meloidogyne incognita. C.I.H. Descriptions of plant-parasitic nematodes.* Set 2, No. 18, St. Albans, UK, C.A.B.

ORTON WILLIAMS, K.J. (1975). *Meloidogyne arenaria. C.I.H. Descriptions of plant-parasitic nematodes.* Set 5, No. 62, St. Albans, UK, C.A.B.

PETERSEN, D.J. & VRAIN, C.V. (1996). Rapid identification of *Meloidogyne chitwoodi* and *M. hapla*, and *M. fallax* using PCR primers to amplifly their ribosomal intergenic spacer. *Fundam. Appl. Nematol.* 19, 601-605.

TRIANTAPHYLLOU, A.C. (1984). Polyploidy in meiotic parthenogenetic populations of *Meloidogyne hapla* and a mechanism of conversion to diploidy. *Rev. Némotol.* 7, 65-72.

TRIANTAPHYLLOU (1985). Cytogenetics, cytotaxonomy and phylogeny of root-knot nematodes. In: *An advanced treatise on Meloidogyne. Vol. I.* pp. 113-126. Eds. J.N. Sasser & C.C. Carter. Raleigh, NC, USA, North Carolina State University Graphics.

VAN DER BEEK, J.G. & KARSSEN, G. (1997). Interspecific hybridization of meotic parthenogenetic *Meloidogyne chitwoodi* and *M. fallax*. *Phytopathology* 87, 1061-1066.

VAN DER BEEK, J.G., FOLKERTSMA, R., POLEIJ, L.M., VAN KOERT, P.H.G. & BAKKER, J. (1997). Molecular evidence that *Meloidogyne hapla*, *M. chitwoodi* and *M. fallax* are distinct biological entities. *Fundam. Appl. Nematol.* 20, 513-520.

VAN MEGGELEN, J.C., KARSSEN, G., JANSSEN, G.J.W., VERKERK, B. & JANSSEN, R. (1994). A new race of *Meloidogyne chitwoodi* Golden *et al.*, 1980? *Fundam. Appl. Nematol.* 17, 93.

WHITEHEAD, A.G. (1968). Taxonomy of *Meloidogyne* (Nematoda: Heteroderidae) with descriptions of four new species. *Trans. Zool. Soc. Lond.* 31, 263-401.

ZIJLSTRA, C., LEVER, A.E.M. & VAN SILFHOUT, C.H. (1995). rDNA restriction fragment length polymorphisms between Dutch isolates of root-knot nematodes. *Nematologica* 19, 355 [Abstr.].

6

ROOT-KNOT NEMATODE PERINEAL PATTERN

DEVELOPMENT: A RECONSIDERATION

'Ausserordentlich deutliche Bilder von Anal- und Genitalspalte erhält
man durch Zerdrüken der trächtigen Thiere auf dem Objectträger. Beide
Spalte sind einander parallel. Die Vulva ist grösser als die Analspalte.'

Carl Müller (1884).

Introduction

The most characteristic morphological feature of the genus *Meloidogyne* is the perineal
pattern, located at the posterior body region of adult females (Fig 1A, E & F). This complex
character comprises the perineum (vulva-anus area), tail terminus, phasmids, lateral lines and
surrounding cuticular striae (Franklin, 1965; Hirschmann, 1985). Different names have been
used for the perineal pattern, like 'anal plate pattern', 'pattern of perineal region', 'posterior
pattern', 'posterior cuticular pattern' and 'fingerprint-like pattern'.
    Müller (1884) observed and illustrated the perineal pattern for the first time (chapter 2).
Chitwood (1949) introduced it as a character for the identification of a small number of root-
knot nematodes. Since then, illustrations of perineal patterns have been added to every species
description. Whitehead (1968) and Jepson (1987) compared in detail morphological
differences between perineal patterns of all known root-knot nematodes. Based on general
pattern morphology, Jepson (1987) divided 50 species into 6 distinct groups. Eisenback
(1985b) predicted that '*The morphology of perineal patterns will probably remain the most
important morphological character used for tentative species identifications*'. He also
mentioned the following reasons why the pattern became the dominant diagnostic character:
for many populations it is quite stable; compared to other useful characters, perineal patterns
are relatively easy to prepare for light microscopic examination; they are relatively large and
two-dimensional; females are more numerous than males and easier to find in infected tissues;
in original descriptions, excellent light micrographs of perineal patterns were presented, other
characters were often illustrated with simple line drawings and too small to point out
effectively the differences among the species.
    However, soon after Chitwood (1949), the first studies on pattern variability appeared.
Allen (1952) noticed morphological pattern variation within a single egg-mass population of
M. incognita, ranging from *M. incognita* to *M. javanica*. He started a single egg-mass culture
of *M. incognita* in 1948 on tomato, after several months the original culture was transferred to
cotton, alfalfa, sugar beet and barley. Three years later he studied perineal pattern variability.
Comparable variation was observed by Netscher (1978) within some Senegalese populations
of root-knot nematode. The studies of Dropkin (1953) and Sasser (1954) on pattern variability

are contradictory with these results. Both workers observed intraspecific variation, but the pattern remained relatively constant within a species and was not changed by host influence. Taylor *et al.*, 1955 proposed a perineal pattern key for the identification of the Chitwood species, including the two subspecies *M. arenaria thamesi* and *M. incognita acrita*. Later both morphological indistinguishable subspecies were correctly synonymized to *M. arenaria* and *M. incognita* respectively (Triantaphyllou & Sasser, 1960; Eisenback & Hirschmann, 1991).

At present most taxonomists accept the intraspecific pattern variation, and recommend for identifications to observe several perineal patterns, including other female, male and second-stage juvenile characters, facultatively added with host differentials or biochemical characters (Esser *et al.*, 1976; Hewlett & Tarjan, 1983; Hirschmann, 1985; Jepson, 1987; Eisenback & Hirschmann, 1991).

Little is known about perineal pattern development, although it could be used to explain the observed morphological variation within or even between species. Therefore literature data on female development and perineal pattern morphology, added with light and scanning electron microscopical observations on different species of European root-knot nematodes, is used to discuss perineal pattern development. Additionally other vulva-anus regions of plant parasitic nematodes with an adult sedentary stage (*Cryphodera brinkmani* Karssen & van Aelst, 1999 and *Nacobbus aberrans* (Thorne, 1935) Thorne & Allen, 1944) were studied, to demonstrate homologies with perineal patterns of root-knot nematodes.

For light and scanning electron microscopical preparation of nematodes, the same methods were used as described in preceding chapters.

From juvenile to adult

A few detailed studies on developmental morphology of root-knot nematodes are available at present. Nagakura (1930), Bird (1959) and Triantaphyllou & Hirschmann (1960) may be considered as classical on this subject. The two latter studies observed the development of *M. javanica, M. incognita* and *M. hapla* on tomato under greenhouse conditions. Other interesting studies on development were published by Inserra *et al.* (1985) and Yu (1995), on *M. chitwoodi* and *M. incognita* respectively. The following developmental description is a compilation on these observations. The time-scheme is focussed on *M. javanica* at 22 °C, according to Bird (1959).

After root penetration of the second-stage juvenile, it starts feeding for about two weeks. During this important period, the juvenile increases in width, the V-shaped primordium moves posteriorly towards the rectum, and the six rectal glands start to increase in length (Fig. 1C). The second-stage juvenile moults within one day, while the anterior part of the stylet is shed, the posterior part (and the metacorpus) breaks down. The tail contents retracts anteriorly, so the posterior body becomes rounded, without tail spike (Fig. 1D). During moulting, the old cuticle is partly reabsorbed by the new underlaying cuticle, as described in detail by Bird & Rogers (1965a). The third-stage juvenile is not feeding (no stylet is present) and there is no change in size, but the rectal glands and the gonads still increase in size. The third stage lasts only for a few hours followed by the third moulting. The third- and also the fourth-stage juveniles are still enclosed by the remnants of the old cuticle(s). The fourth stage is also non-feeding, without any size change. During this stage, of about one day, the vagina and the uterus start to differentiate, the gonads and the rectal glands expand rapidly (Fig. 1D) and the metacorpus reappears.

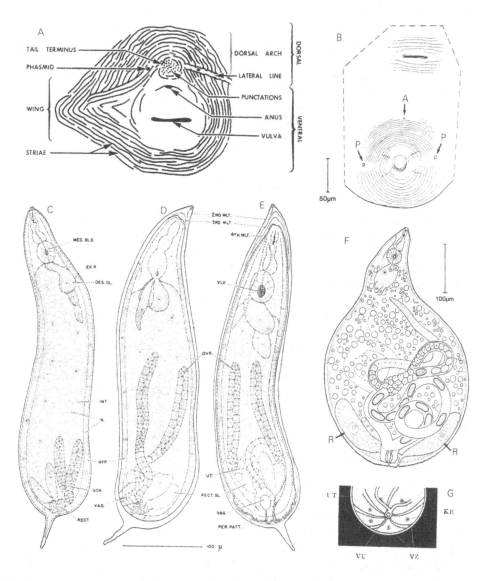

Fig. 1. A: *Meloidogyne* perineal pattern diagram; B: Perineal area of *Meloinema kerongense* (A= anus, P= phasmid); C-E: Female *M. incognita* development (C= swollen J2 before second moult, D= fourth stage juvenile, E= adult shortly after fourth moult); F: Cross-section *M. microcephala* female (R= rectal gland); G: Female posterior end with vulva and six rectal gland nuclei. (A: after Eisenback *et al.*, 1981; B: after Choi & Geraert, 1974; C-E: after Triantaphyllou & Hirschmann, 1960; F: Hirschmann, 1985; G: Nagakura, 1930).

The complete period of moulting, from the second to the last moult, occurs within two to three days. Just after the last moult, while the young adult female is still enclosed by the old cuticle layers, the stylet, vulva and the perineal pattern appear. The new cuticle continues to increase in thickness particularly in the posterior region (Bird & Rogers, 1965a). The nematode starts feeding again and increases extremely rapid in width, after 9 to 10 days followed by the first gelatinous matrix formation (produced by the rectal glands), while the posterior part of the body breaks through the root surface. Two days later, the female reaches maximum size and starts with egg-laying and continues with the production of a large amount of gelatinous matrix.

The old cuticles

The young adult female is still enclosed by the remnants of the old cuticles. Triantaphyllou & Hirschmann (1960), reported about the old cuticle layers '*The cuticles of the previous molts disappear shortly afterwards*'. They simply disappear, but how and where? According to Franklin (1965) '*In the female the cast cuticles are pushed aside and feeding recommences.*' So the old layers are pushed aside because of the sudden increase in width? Bird & Rogers (1965a) reported however '*This layer* (external cortical layer, epicuticle or cortical layer) *is not reabsorbed but is retained and often seen* (with the transmission electron microscope) *as a loose sheath around parts of adult specimens*'. These are some of the very rare reports on old culticle layers in young adult *Meloidogyne* females. With a few exceptions, I have never observed these old cuticle layers with the light microscope (LM).

During the redescription of *M. maritima* (Jepson, 1987) Karssen, van Aelst & Cook (1998), we observed (LM) some 'loose cuticle parts' in young TAF fixed females, but did not recognize it as the remnants of the old cuticle layers, until we started with scanning electron microscopical (SEM) observations of young egg-laying females. The first SEM observations of the female S-E pore (Fig. 2C) showed that some of the pores and their surrounding annulations were 'covered', while others were not (Fig. 2D, see also 5A). Further observations in the anterior region, clearly showed the remnants of the old cuticle layer (Fig. 2A & B). We observed these layer only in the neck region of swollen females. *M. maritima* females induces very small galls, and only the neck region of adult females is embedded in the root tissue (chapt. 4). These observations indicate that it is likely that the old culticle layer is pushed aside by the rapid grow of the female, as Franklin (1965) speculated. If so, one would expect to find a completely covered female just after the last moult, before width increase.

For this reason young egg-laying females of *M. kralli* Jepson, 1983 were carefully isolated. Only females were selected with a small gelatinous matrix, most of them filled with only a few eggs. This species also induces very small galls, as in *M. maritima*. Just after the last moult one will find the young adult females partly embedded in root tissue with the posterior region clearly visible on the root surface. The young (SEM) observed females were indeed almost completely covered with the old cuticle layer (Fig. 3A). Near mid-body cracks were observed, and in the posterior region folded cuticle parts (Fig. 3C & D), indicating the 'pushed aside' effect.

Another rather surprising observation was the abundant occurrence of an unknown rod-shaped bacterium on the old cuticle layer (Fig. 3C-D). Coincidentally a comparable bacterium was observed on *Cryphodera brinkmani* (Fig. 7). Although the 'pushed aside' effect is likely, it doesn't explain what finally happens to the old culticle layers. Hypothetically the observed bacterium could play a key role in the ultimate break down of these old layers.

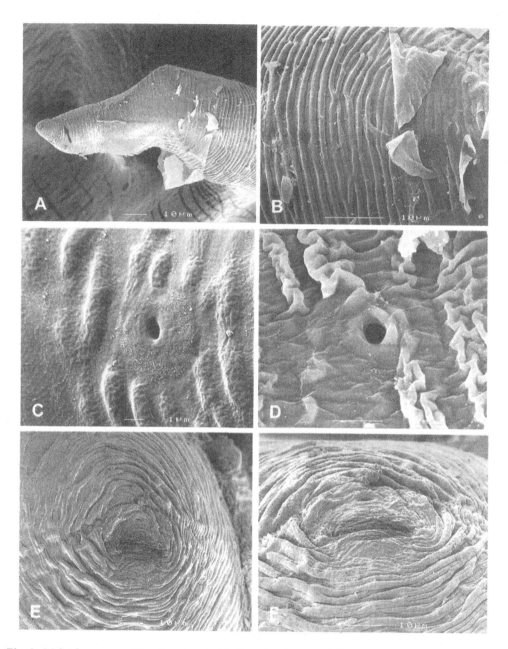

Fig. 2. *Meloidogyne maritima* females. A: Neck region (arrow= S-E pore); B: Cuticle detail in neck region; C: S-E pore with old cuticle layer; D: S-E pore; E: Perineal pattern; F: Perineal pattern in side view.

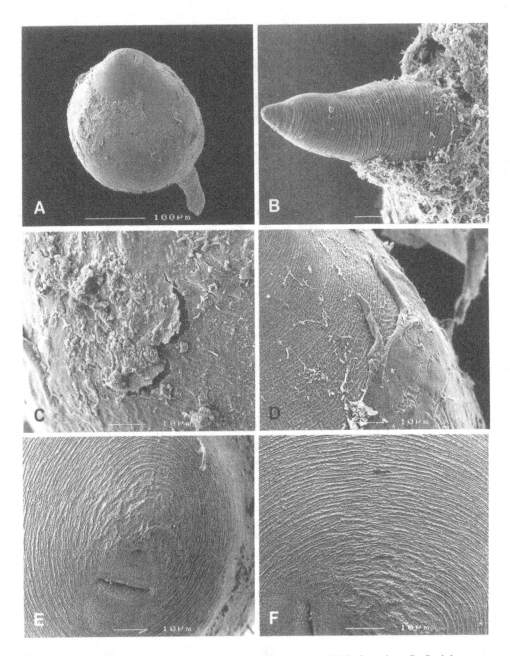

Fig. 3. *Meloidogyne kralli* females. A: Habitus, upside-down; B: Neck region; C: Cuticle near mid-body; D: Cuticle in posterior region; E: Perineal pattern; F: Perineal pattern lateral region.

Perineal pattern

Basic pattern

The basic features of the perineal pattern, i.e. vulva, anus, phasmids, tail terminus, lateral lines and annulations, are already present in the early adult stage, before the female starts to increase in width. The entire female body is regularly annulated at this stage (Fig. 3). However around the vulva and the anus the incisions forming the annuli are closer together and deeper in most adults, and named striae. They form the most characteristic part of the perineal pattern (Eisenback, 1985b). In some rare species, like *M. kralli*, there is no difference between the regular body annulations and pattern striae in adult females (Fig. 3E). Only at the lateral sides of the vulva one may observe some deviating annulations (Fig. 3F).

First feeding stage (J2)

Tyler (1933) observed already variation in rate of development in juveniles and related this with their nutrition. Triantaphyllou & Hirschmann (1960) reported '*A striking feature in the development of M. incognita seems to be the great variation in degree of development among individual larvae. This variation may result from differences in rate of development during the second larval stage when the nematodes are feeding*'. And explained this by '*Crowding and position of each individual larva in the root tissues may be two factors affecting the amount of food available to the nematode*'. If the J2 feeding state and development are indeed strongly correlated, it is likely that this effects also later stages of development, in particular the non-feeding stages (J3 & J4). Initial differences in development caused by the J2 feeding state may be partly diminished by the second feeding stage (the adult). It is hypothesized that these initial differences effect perineal pattern development. Slightly compressed body annuli are already present in this stage, and observed during this study, near the (swollen) anus region of feeding *M. hapla* second-stage juveniles.

Pre-adult

In the majority of the *Meloidogyne* species, the perineal pattern is located in a central posterior position. How is this possible? Müller (1884) described it as '*Gegenseitige Lage und Grössenverhältniss zeigt Fig. 1, Taf. III* (chapt. 2, Fig. 1) *hier heben sich zugleich die eng an einander gepressten..*'. Nagakura described an interesting observation '*Der after welcher anfänglich nähe der Geslechtsöffnung auf der Bauchseite liegt, ist im letzten Stadium in der Nähe der Vulva auf der Rückenfläche verlagert*'. This is an important remark, the vulva is apparently invaginated just above the anus on the ventral side; both shift posteriorly. Finally the vulva is in a posterior position, the anus in a sub-dorsal position and the basic perineal pattern is present. I was able to reconstruct this process, while observing *M. hapla*, *M. duytsi* and *M. fallax* pre-adults. The anus was in this stage not clearly covered by a cuticular fold, as described for the adult females of these species (Karssen & van Hoenselaar, 1998). Triantaphyllou & Hirschmann (1960) reported indirectly about this process '*In young females the body contents contract and separate from the cuticle*', and they clearly illustrated it (Fig. 1D-E). Strangely I have not observed the vulva-anus shift in *M. kralli*, the vulva and anus are fixed in a sub-ventral position, even in older egg-laying adult females (Fig. 3A). The vulva-anus shift may be marked as a (second) potential cause of perineal pattern variability.

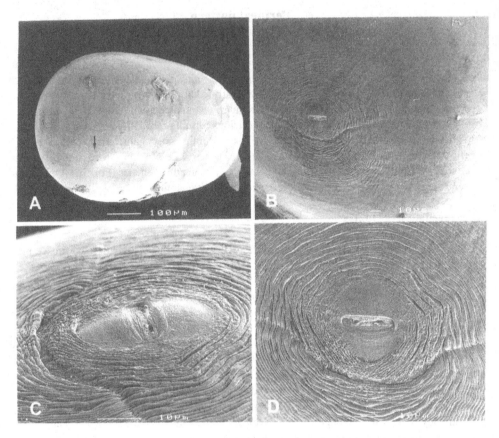

Fig. 4. *Meloidogyne hapla* female. A: Habitus (arrow= lateral line); B-D: perineal patterns.

Second feeding stage (adult)

After the female starts feeding again, she increases in width. '*Between the twentieth and twenty-seventh days growth is extremely rapid. The nematodes more than quadruple their cross sectional area*' (compared to J3), according to Bird (1959). '*The mean diameter of nematode females increased 16-30 fold, i.e. from 16-μm J2 and 30-μm J3 to the 492-μm adults, between 8 and 40 d after inoculation*', after Yu (1995). This is the reason why the body annulations and lateral lines in most *Meloidogyne* species simply disappear between the neck region and the posterior end (Fig. 4A). Or as Eisenback (1985a) described '*Body annulations in females mark the anterior portion of the body, including all of the neck. As they proceed posteriorly, the annulations become shallower and farther apart until they disappear completely. These annulations gradually reappear on the posterior portion of the body*'.

There are differences in the final female body shape between *Meloidogyne* species (Whitehead, 1968 and Jepson, 1987), but this character is only useful to cluster the species roughly into six groups (Eisenback & Hirschmann, 1991).

Size increase

A rather logical question is, if the extreme size increase effects the basic perineal pattern shape? The above described size increase may suggest that the perineal pattern is situated on an untouched posterior position. However this is not correct, the perineal pattern of very young females still enclosed in the juvenile cuticle is markedly different from mature females (Triantaphyllou & Sasser, 1960). While in young females the pattern lines appear as fine striae, slightly thicker than regular body annulations, in mature females one will observe coarse striae and folds. I have observed this in the following *Meloidogyne* species: *M. ardenensis*, *M. duytsi*, *M. fallax* and *M. maritima*, but not in *M. kralli*. During the mid-body width increase, the perineal region changed in the observed species from hemispherical to plain (except *M. kralli*). Also the anus is effected, a small cuticular fold appears above the anus and covers it partly in most species, so the gelatinous matrix is extruded towards the vulva. (Fig. 4D & 5B-D).

It is clear that the morphological change of the perineal pattern correlates with the size increase of the adult female. This is the main reason why for taxonomical studies and if possible also for routine identifications, one should observe females of the same age. The young egg-laying female stage is preferable, as it is easy to recognize because of the small corresponding egg-mass. And more important, at this stage the female reaches maximum size (Bird, 1959).

Rectal glands

Simultaneously with the female size increase, the rectal glands expand rapidly. Just before the female starts with egg-laying, the six rectal glands reaches maximum size and start their activity (Triantaphyllou & Hirschmann, 1960; Fig. 1F). They produce a large amount of gelatinous matrix through the anus. Each gland is pyriform, about 100 μm long and 25 μm width. The six glands are closely situated to the hypodermis and equally distributed within the posterior body (Maggenti & Allen, 1960; Bird & Rogers, 1965b; Fig 1G). See Geraert (1994), for an interesting discussion on the origin and function of the gelatinous matrix.

Is the body width increase responsible for the appearance of coarse striae, as proposed by Triantaphyllou & Sasser (1960)? Or do the expanding rectal glands play a role in the posterior cuticle stretching, or is it a combination of both? Within the genus *Meloidogyne* there is one group of species with a more elongated body shape (= less width increase), the pattern region is still hemispherical. The perineal pattern appears as if it is situated on a protuberance or as if the perineum is raised. These species, most of them placed in the former genus *Hypsoperine*, are: *M. acronea*, *M. aquatilis*, *M. californiensis*, *M. graminicola*, *M. graminis*, *M. ichinohei*, *M. kongi*, *M. kralli*, *M. lini*, *M. marylandi*, *M. mersa*, *M. oryzae*, *M. ottersoni*, *M. propora*, *M. salasi*, *M. sasseri*, *M. spartinae* and *M. triticoryzae*. All these species have a perineal pattern with fine striae, except for *M. graminis*, *M. marylandi* and *M. sasseri* with coarse striae. However most of them produce a gelatinous matrix, therefore the expanding rectal glands play not an important role in stretching of the posterior cuticle in these species. It is more likely that the observed fine striae formation is the result of less body width expansion.

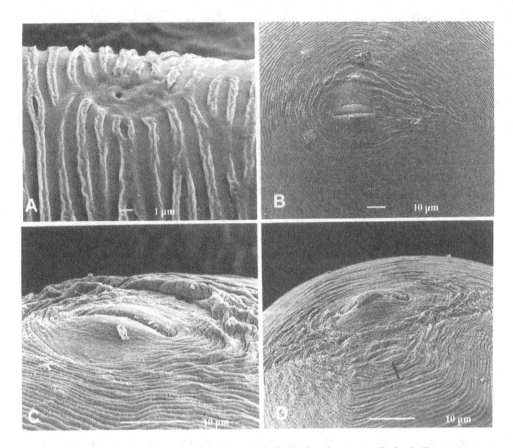

Fig. 5. *Meloidogyne duytsi* female. A: S-E pore; B-D: Perineal patterns. B: Including 'wing';
C: Ventral view; D: Dorsal view (arrow= 'tail terminus').

Human factor

An intriguing question is, are perineal pattern descriptions it'self a source of variability?
Therefore the eight most important descriptions of the well-known species *M. hapla*, are
herein listed.

*'Transverse striae in perineal region extremely smooth, tending to form a rounded hexagon,
lines rather continuous posterior to anus, tending to be parallel and more widely spaced than
elsewhere, sometimes with slight shoulder at level of phasmids, but without continuing lateral
incisures. Pattern of these striae with no post-anal whorl, sometimes with one right or left
lateral loose loop, sometimes with paired lateral loops.'* (**Chitwood, 1949**)

'The lateral lines may be marked only by slight irregularities in the striae, or the striae of the dorsal and ventral sectors may meet at an angle along the lateral lines. The arch is low. Near the tip there is often an area with distinct stippling or punctation which is not found in other species. The ventral striae are sometimes extended laterally to form wings on one or both sides. The striae in both sectors are smooth or slightly wavy.' (**Taylor et. al., 1955**)

'Wölbung niedrig und rundlich, gelegentlich auch dorsal abgeflacht und im Scheitelpunkt etwas eingesunken. Seitenlinien sind nur durch Unregelmässigkeiten oder kurze Unterbrechungen der Kutikularstreifen angedeutet. Die Streifen selbst sind deutlich, jedoch durchweg nicht kräftig entwickelt, abgesehen vielleicht von den zentral verlaufenden. Alle Linien sind meist schwach wellenförmig ausgebildet. Die Streifen des dorsalen und ventralen Sektors können an der Seitenlinie einseitig oder auf beiden Seiten schwingenartig ausgebuchtet sein.' (**Goffart, 1957**)

'Posterior cuticular pattern roughly circular, the smooth striae forming a low arch; lateral fields usually not marked but the striae at level of the vulva may form a loop or wing which the dorsal striae meet almost at right angles. The tail end has few striae but there is often a stippled area. The anus is usually covered by a fold of cuticle.' (**Franklin, 1965**)

'Posterior cuticular pattern mostly composed of closely spaced smooth or slightly wavy striae, dorsal arch low rounded, lateral lines present, stippling always present, either concentrated in area between anus and tail terminus or more diffuse over inner part of pattern, some forking of striae at the lateral lines and occasionally 'fringing' striae observed at posterior ends of the lateral lines.' (**Whitehead, 1968**)

'Posterior cuticular pattern roughly circular, composed of closely spaced smooth or slightly wavy striae. Dorsal arch low. Lateral fields may be unmarked, may be marked only by slight irregularities in the striae, or dorsal and ventral striae may meet at a slight angle along the fields. Some forking of striae at the lateral fields may also occur. In some cases ventral striae may extend laterally on one or both sides to form "wings" which the dorsal striae meet almost at right angels. Tail with few striae but distinct punctations forming a stippled area between the anus and tail terminus. Sometimes the stippling may be more diffuse over the inner part of the pattern.' (**Orton Williams, 1974**)

'Perineal patterns are characterized by their overall rounded hexagonal to flattened ovoidal shape, very fine striae, and subcuticular punctations in the smooth tail terminus area. The dorsal arch is usually low and rounded, but may be high and squarish. Lateral ridges are absent but the lateral fields are marked by irregularities in the striae. The dorsal and ventral striae often meet at an angle, and the striae are smooth to slightly wavy. Some patterns may form wings on one or both lateral sides.' (**Eisenback, 1985b**)

'Perineal pattern circular or oval; dorsal arch low to medium high, apex broadly rounded, sometimes slightly squarish, dorsolateral indentations shallow or absent, areas above lateral fields often bulged outwards, inner striae form a low, rounded pattern above tail terminus; lateral fields indicated by forking of striae, or the dorsal and ventral striae meet at an angle, forming a single recessed line in each field, or a trough containing short vertical striae forms in each field around and outside phasmids; tail terminus with surface punctations which are *usually enclosed by lines extending between phasmids; anal flap often ventrolaterally together with slightly flattened rectal lining as a wide perineal ridge; rectal punctations usually indistinct, arranged close to rectal lining; striae of ventral arch often form a wing on one or both sides of the pattern.'* (**Kleynhans, 1991**)

Although all these taxonomists described a *M. hapla* perineal pattern, and mentioned the most important features as a low dorsal arch, presence of wings and punctations, a lateral field and fine striae, the differences (or human factor) in these descriptions are striking. Particularly the striae shape and lining is variable. I consider the relatively simple pattern description of Franklin (1965) as the clearest one. Franklin (1972) mentioned also the problem of pattern descriptions *'The perineal pattern has two great disadvantages, that it is difficult to define and it is often very variable within species. Attempts have been made to find a simple and reliable way of describing perineal patterns, but nothing suitable has been developed....'* She indirectly recommended to reduce the human factor by using photographs instead of drawings *'Good photographs give a truer picture of a specimen.'* When describing perineal patterns it is better to avoid variable striae details as *forking, slight irregularities, short vertical, fringing* etc., and to focus more on pattern details as lateral lines, punctations, wings and dorsal arch shape.

Measurements

Pattern measurements as vulva length, vulva-anus distance, dorsal arch height and inter-phasmidial distance are, because of variability, considered as of little value for comparing *Meloidogyne* species (Jepson, 1987; Whitehead, 1968). However they may be useful, if we know all the factors influencing pattern development, to correlate   detailed measurements of the second-stage juvenile tail with pattern measurements. Tail measurements as anus-phasmid distance, anus-anterior hyaline tail part length, body width at anus and phasmid level and anterior rectum-anus distance may be good candidates (after correction for factors influencing pattern development) to develop a (predictive) model for pattern development.

'Tail terminus'

The tail terminus is located between the phasmids and just above the anus (Fig. 1A). It is partly a remnant of the second-stage juvenile tail, during the second moulting the tail contents retract anteriorly as described before. In some species it is clearly visible with the scanning electron microscope (SEM) as a small rounded structure (Fig. 5C), or as a whorl (Fig. 3E), or as a small compressed stripe (Fig. 2F). In general this structure is named tail terminus or tail tip, suggesting a correlation with the second-stage juvenile tail terminus. However, as observed several times during this study, not the complete tail contents retract during moulting. Only the tail contents above the hyaline tail terminus retract anteriorly.

In conclusion, the name 'tail terminus' within the perineal pattern is incorrect or at least misleading. Therefore the name *tail remnant* or *posterior tail remnant* is proposed herein as more correct.

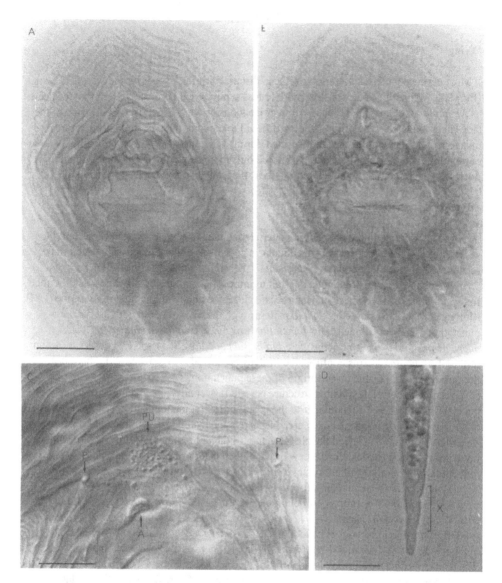

Fig. 6. *Meloidogyne ardenensis* (A & B); *M. hapla* (C & D). A: Perineal pattern; B: Same pattern, subcuticular; C: Perineal pattern detail (P= phasmid, Pu= punctations, A= anus); D: J2 posterior part (X= unclear deliminated hyaline part).

Discussion

'Wings'

Most perineal pattern structures have been discussed so far, except for the so called 'wings' (Fig. 1A). While most species have more or less symmetrical perineal patterns, some species as *M. hapla* and *M. duytsi* regularly form striae loops or wings, in most cases on one of the sides near the vulva. The origin is rather difficult to explain, it is not some kind of rare or unstable pattern deviation. Are they formed during (asymmetrical) vulva induction? Is there a correlation with the slightly bending of the J2 tails in both species? Or is there something present as asymmetrical female swelling? It's all speculative and needs more research.

Subcuticular structures

In fixed and stained perineal pattern material, one may observe a 'dark area' around the vulva (Fig. 6A). When focussed subcuticular it appears as a 'dark ring'. This structure is related with the vagina, probably a remnant of the vagina wall or vagina muscles.

Another subcuticular structure is the 'tail terminus' punctations or stipplings, as described for *M. hapla* (Fig. 1A). These punctations are clearly visible with the light microscope (LM). Mulvey *et al.* (1975) studied *M. hapla* patterns with the SEM, but did not observe any punctations. They concluded '*This punctation was not evident on scanning electron micrographs and is therefore considered to be subcuticular.* SEM and LM observations on different *M. hapla* patterns confirms the Mulvey *et al.* study (Fig. 5D & 6C). There is no literature record available on the origin of these punctations.

Is it possible to correlate it with a structure present in second-stage juvenile tails? The punctations are located subcuticular to the posterior tail remnant in patterns. Theoretically these punctations originate from an area just above the hyaline tail terminus of second-stage juveniles. Bird (1979) studied the ultrastructure of the tail region of *M. javanica* second-stage juveniles in detail. He described a new organ in the posterior region of the tail: '*Just anterior to the tail tip (= hyaline tail terminus) a space lined by the fibrillar part of the basal layer occurs and in this space and closely opposed to the basal layer of the cuticle is a sensory structure made up of numerous branching nerve processes. I have called this structure the caudal sensory organ. The nerve processes originate from several axons which together probably constitute the median caudal nerve. This large mechanoreceptor is approx 5-10 μm in length. So far as I am aware this is the first time that a sensory structure of these dimensions has been described in the tail tip of nematodes.*' Is there a correlation between the caudal sense organ (CSO) and perineal pattern punctations? Are the punctations related to the nerve processes? If the CSO is widespread within the genus *Meloidogyne* one would expect to find much more species with pattern punctations, however it is only described for *M. hapla*. Probably the CSO is correlated with another typical *M. hapla* structure. The anterior hyaline tail terminus is not clearly deliminated and runs deep into the tail terminus (x in Fig. 6D), the x-part also retracts during moulting. It is not unlikely that the CSO also runs deep into the hyaline tail terminus of *M. hapla*. A transmission electron microscopical (TEM) study of this tail part would be useful to clearify the origin of the perineal pattern punctations.

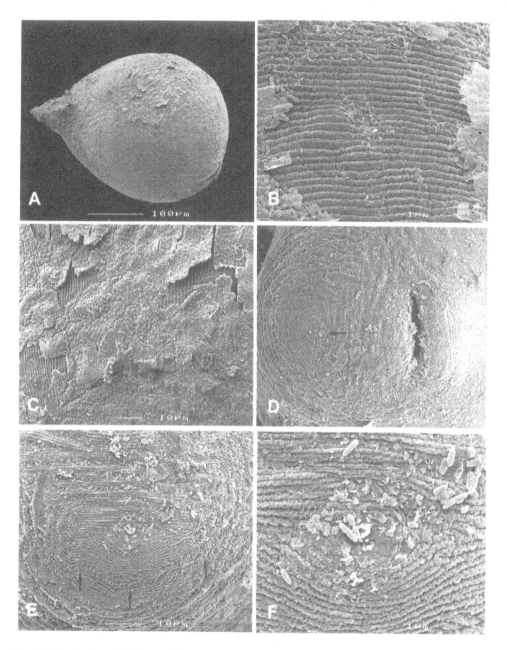

Fig. 7. *Cryphodera brinkmani* female. A: Habitus; B & C: Cuticle and old cuticle layer; D: Vulva-anus region (arrow= anus); E: Posterior region (arrows= lateral line); F: Anus.

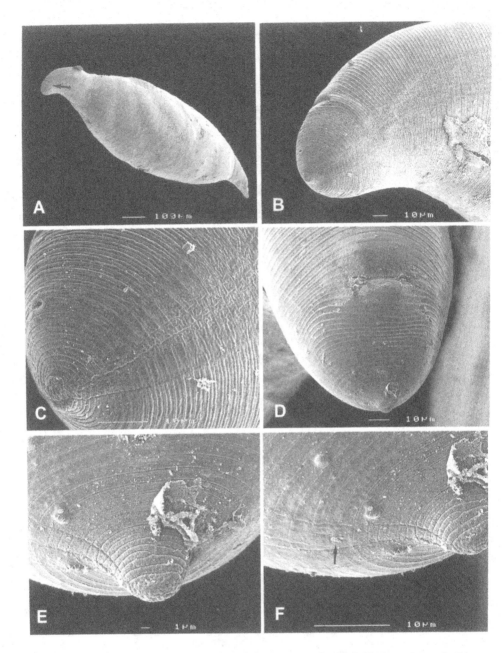

Fig. 8. *Nacobbus aberrans* female. A: Habitus (arrow= posterior end); B: Posterior end (side view); C: Tail (side view); D: Vulva-anus region; E: Tail tip; F: Tail tip (arrow= phasmid).

Image position

After Chitwood (1949) perineal patterns have been depicted with the dorsal side upwards and the ventral side down (Fig. 1A). However is this image position correct? In most biological illustrations, individuals are depicted with their anterior part upwards and their posterior part down. If one illustrates a part of a nematode body, for example a tail of a *Pratylenchus* male, it is depicted with the tail tip down. Also in sedentary adult female nematodes where the vulva has been shifted towards the tail, the vulva-tail tip region is logically illustrated with the vulva upwards and the tail tip down. *Meloinema kerongense* Choi & Geraert (1974), an unusual plant parasitic nematode placed in the family Heteroderidae (subfamily Nacobboderinae) has not a real perineal pattern, but all landmarks as a subterminal vulva, anus, tail tip and phasmids are present (Fig. 1B). The other Nacobboderinae genera, *Nacobbodera* and *Bursadera*, have also comparable vulva-anus-tail tip regions.

Another Heteroderidae example (subfamily Heteroderinae), with the vulva and anus more closer, is *Cryphodera brinkmani*. In Fig. 7D the subterminal vulva-anus region is placed on one of it's lateral sides with the vulva and the anus on an imaginary horizontal line, to show the effect of an incorrect depicted body part. In Fig. 7E it is correctly depicted, anus upwards and tail tip with lateral line remmants down. However from the other non-cyst forming Heteroderinae with a terminal to subterminal vulva and small vulva-anus distance as *Hylonema, Bellodera, Ekphymatodera, Atalodera, Rhizonema* and *Sarisodera*, the posterior regions (if studied) have been depicted in several ways. Another situation was recorded for the cyst forming Heteroderinae as *Heterodera, Cactodera, Globodera* and *Punctodera* where the vulva-anus regions are almost equally depicted with their vulva upwards, anus down, or the other way round.

Also a nematode genus outside the Heteroderidae, but with a swollen adult female stage and a subterminal vulva, was SEM studied posteriorly. The false root-knot nematode *Nacobbus aberrans s. l.* (Nematoda: Pratylenchidae) vulva-anus region is also comparable with *Meloinema kerongense*. Although not a real perineal pattern also here are all the basic structures present, as vulva, anus, tail tip, phasmids and lateral lines (Fig. 8), and depicted correctly in Fig. 8D.

The perineal pattern of the genus *Meloidogye* is comparable with the vulva-anus region of *Meloinema, Cryphodera, Nacobbus* and others, except for the vulva-anus distance. Within *Meloidogyne* this distance is very small, so small that only a few rare species as *M. coffeicola, M. decalineata, M. indica, M. nataliei* and *M. propora* (= pattern group 2 after Jepson, 1987) have some continuous striae between the vulva and anus. Within *Meloidogyne* the vulva is also induced above the anus on ventral side, but shifts posteriorly in most species, as discussed earlier. In Fig. 4D the perineal pattern of *M. hapla* is depicted in the same way as the vulva-anus region of *Meloinema kerongense* in Fig. 1B. Conclusion: it is rather surprising that the perineal pattern has been incorrectly depicted upside-down for fifty years.

Epilogue

At present we may recognize the following origins for perineal pattern variation:
1) Developmental factors as the first feeding stage, vulva-anus shift and second feeding stage.
2) Human factors as pattern interpretation or description and the observation of patterns of different age.

The perineal pattern is an interesting character or character complex for developmental studies. The root-knot nematode nematode is also a useful organism for such studies, as it is easy to culture and the life-cycle is short. Futher observations, in particular with the transmission electron microscope and on genetical level, would be helpful to unravel the true nature of the developmental factors in relation to perineal pattern origin.

Literature

ALLEN, M.W. (1952). Observations on the genus *Meloidogyne* Goeldi, 1887. *Proc. Helminth. Soc. Wash.* 19, 44-51.

BIRD, A.F. (1959). Development of the root-knot nematodes *Meloidogyne javanica* (Treub) and *Meloidogyne hapla* Chitwood in tomato. *Nematologica* 4, 31-42.

BIRD, A.F. (1979). Ultrastructure of the tail region of the second-stage preparasitic larva of the root-knot nematode. *Int. J. for Parasitology* 9, 357-370.

BIRD, A.F. & ROGERS, G.E. (1965a). Ultrastructure of the cuticle and its formation in *Meloidogyne javanica*. *Nematologica* 11, 224-230.

BIRD, A.F. & ROGERS, G.E. (1965b). Ultrastructure and histochemical studies of the cells producing the gelatinous matrix in *Meloidogyne*. *Nematologica* 11, 231-238.

CHITWOOD, B.G. (1949). Root-knot nematodes, part I. A revision of the genus *Meloidogyne* Goeldi, 1887. *Proc. Helminth. Soc. Wash.* 16, 90-104.

CHOI, Y.E. & GERAERT, E. (1974). Description of *Meloinema kerongense* n. g. n. sp. (Nematoda: Meloidogynidae) from Korea. *Nematologica* 19, 334-341.

DROPKIN, V.H. (1953). Studies on the variability of anal plate patterns in pure lines of *Meloidogyne* spp. the root-knot nematode. *Proc. Helminth. Soc. Wash.* 20, 32-39.

EISENBACK, J.D. (1985a). Detailed morphology and anatomy of second-stage juveniles, males and females of the genus *Meloidogyne* (Root-Knot nematodes). In: *An advanced treatise on* Meloidogyne. *Volume I, Biology and Control.* pp. 47-77. Ed. J.N. Sasser & C.C. Carter. Raleigh, USA, North Carolina State University Graphics.

EISENBACK, J.D. (1985b). Diagnostic characters useful in the identification of four most common species of root-knot nematodes (*Meloidogyne* spp.). In: as Eisenback (1985a). pp. 95-112.

EISENBACK, J.D. & HIRSCHMANN, H. (1991). Root-Knot nematodes: *Meloidogyne* species and races. In: *Manual of agricultural nematology.* pp. 191-274. Ed. W.R. Nickle. New York, Marcel Dekker, inc.

EISENBACK, J.D., HIRSCHMANN, H., SASSER, J.N. & TRIANTAPHYLLOU, A.C. (1981). *A guide to the four most common species of root-knot nematodes* (Meloidogyne *species), with a pictorial guide.* A coop. Publ. Depts. Plant Pathol. and Genetics and U.S. Agency International Dev., Raleigh, NC.

ESSER, R.P., PERRY, V.G. & TAYLOR, A.L. (1976). A diagnostic compendium of the genus *Meloidogyne* (Nematoda: Heteroderidae). *Proc. Helminth. Soc. Wash.* 43, 138-150.

FRANKLIN, M.T. (1965). *Meloidogyne*-Root-Knot Eelworms. In: *Plant Nematology*. pp. 59-88. Ed. J.F. Southey. London, H.M.S.O.

FRANKLIN, M.T. (1972). The present position in the systematics of *Meloidogyne*. *OEPP/EPPO Bull.* 6, 5-15.

GERAERT, E. (1994). The origin of the gelatinous matrix in plant parasitic nematodes. *Nematologica* 40, 150-154.

GOFFART, H. (1957). Bemerkungen zu einigen arten der gattung *Meloidogyne*. *Nematologica* 2, 177-184.

HEWLETT, T.E. & TARJAN, A.C. (1983). Synopsis of the genus *Meloidogyne* Goeldi, 1887. *Nematropica* 79-102.

HIRSCHMANN, H. (1985). The genus *Meloidogyne* and morphological characters differentiating its species. In: as Eisenback (1985a). pp. 79-93.

INSERRA, R.N., VOVLAS, N., O' BANNON, J.H. & GRIFFIN, G.D. (1985). Development of *Meloidogyne chitwoodi* on Wheat. *J. Nematol.* 17, 322-326.

JEPSON, S.B. (1987). *Identification of root-knot nematodes* (Meloidogyne *species*). Wallingford, UK, C.A.B. International.

KARSSEN, G. & VAN HOENSELAAR, T. (1998). Revision of the genus *Meloidogyne* Göldi, 1892 (Nematoda: Heteroderidae) in Europe. *Nematologica* 44, 713-788.

KARSSEN, G. & VAN AELST, A. (1999). Description of *Cryphodera brinkmani* n. sp. (Nematoda: Heteroderidae), a parasite of *Pinus thunbergii* Parlatore from Japan, including a key to the species of the genus *Cryphodera* Colbran, 1966. *Nematology* 1 (2), in print.

KLEYNHANS, K.P.N. (1991). *The root-knot nematodes of South Africa*. Techn. comm. Dep. of Agric. Develop. No. 231.

MAGGENTI, A.R. & ALLEN, M.W. (1960). The origin of the gelatinous matrix in *Meloidogyne*. *Proc. Helminth. Soc. Wash.* 27, 4-10.

MÜLLER, C. (1884). Mittheilungen über die unseren Kulturpflanzen schädlichen, das Geschlecht *Heterodera* bildenden Würmer. *Landw. Jahrb.* 13, 1-42.

MULVEY, R.H., JOHNSON, P.W., TOWNSHEND, J.L. & POTTER, J.W. (1975). Morphology of the perineal pattern of the root-knot nematodes *Meloidogyne hapla* and *M. incognita*. *Canadian J. Zoology.* 53, 370-375.

NAGAKURA, K. (1930). Ueber den Bau die Lebensgeschichte der *Heterodera radicicola* (Greef) Müller. *Japanese J. Zoology* 3, 95-160.

NETSCHER, C. (1978). Morphological and physiological variability of species of *Meloidogyne* in West Africa and implications for their control. *Mededelingen Landbouwhogeschool Wageningen* 78-3, 1-46.

ORTON WILLIAMS, K.J. (1974). *Meloidogyne hapla. C.I.H. Descriptions of plant-parasitic nematodes.* Set 3, No. 31, St. Albans, UK, C.A.B.

SASSER, J.N. (1954). Identification and host-parasite relationships of certain root-knot nematodes (*Meloidogyne* spp.). *Bull. Maryland Agric. Exp. Stn.* No. A-77, 1-30.

TAYLOR, A.L., DROPKIN, V.H. & MARTIN, G.C. (1955). Perineal patterns of the root-knot nematodes. *Phytopathology* 45, 26-34.

TRIANTAPHYLLOU, A.C. & HIRSCHMANN, H. (1960). Post-infection development of *Meloidogyne incognita* Chitwood, 1949 (Nematoda: Heteroderidae). *Ann. Inst. Phytopathol. Benaki* 3, 3-11.

TRIANTAPHYLLOU, A.C. & SASSER, J.N. (1960). Variation in perineal patterns and host specificity of *Meloidogyne incognita. Phytopathology* 50, 724-735.

TYLER, J. (1933). Reproduction without males in aseptic root cultures of the root-knot nematode. *Hilgardia* 7, 373-388.

WHITEHEAD, A.G. (1968). Taxonomy of *Meloidogyne* (Nematoda: Heteroderidae) with descriptions of four new species. *Trans. zool. Soc. Lond.* 31, 263-401.

YU, M.H. (1995). Root-knot nematode development and root gall formation in sugarbeet. *J. of Sugar Beet Research* 32, 47-58

7

SUMMARY

The plant-parasitic nematode genus
Meloidogyne Göldi, 1892 (Tylenchida)
in Europe

Root-knot nematodes (genus: *Meloidogyne* Göldi, 1892) represent a relatively small but important group of plant parasitic nematodes. World-wide more than eighty species have bee described, while twenty species have been detected in Europe so far. This thesis includes a historical review on the genus, followed by a revision of the European species, and complete with a study on the development of one of the most characteristic morphological structures within the genus: the perineal pattern.

Historical notes

The history of the root-knot nematodes is marked by three distinct periods. In the early period (1855-1878) the causal relation of the nematode to a plant disease known as root-kno' was proved. This little soil inhabiting nematode induces galls or knots on many plant roots. The root parasite was reported for the first time by the Englisman Berkeley in 1855, soon followed by other reports from Europe and other continents. During the early period another important root parasite was discovered (1859) and described (1871) as *Heterodera*. This genus is related but different from root-knot nematodes, as it induces no root galls and the adult females are transformed into a cyst.

During the middle period (1879-1948), root-knot nematodes and cyst nematodes were in gerneral named as one genus, despite the clearly documented differences by Marcinowski (1909) and other workers. It was the Frenchman Cornu who described the first root-knot nematode, as *anguillula marioni*, from Europe in 1879. It was rather poorly described and n type material was deposited, today this species is marked as a *species inquirenda*. While in the middle period the number of accepted *Heterodera* species increased, root-knot nematode were treated as one species and named *Heterodera radicicola* (1879-1931) and *Heterodera marioni* (1932-1949) respectively. We may consider the period 1879-1948 as a taxonomical confusing period on genus and species level for root-knot nematodes. On the other hand it w an exciting period of great discoveries. Like the first perineal pattern observations by Müllei (1884), or the detailed morphological study of Nagakura (1930). Although the root-knot nematodes were treated as one species, some of the agricultural most important were described during this period, like *Meloidogyne javanica* (Treub, 1885), *M. exigua* (Göldi, 1892), *M. arenaria* (Neal, 1889) and *M. incognita* (Kofoid & White, 1919). The following period (1949-1998) was carefully introduced by the publications of Christie *et al.* in 1944 ar 1948, when they proved the existence of 'species' within the genus.

Root-knot nematode taxonomy really started in 1949 when Chitwood published his revisic on the genus. He re-erected the genus *Meloidogyne* and redescribed a number of species. Th small publication changed taxonomy thoroughly. The species number increased rapidly and detailed studies on morphology, isozymes, cytogentics and recently DNA appeared in this period. However the number of critical revisions or monographs on root-knot nematodes is surprisingly limited.

Revision

The species status of the European root-knot nematodes (*Meloidogyne* Göldi, 1892) was studied. Types of *M. arenaria, M. artiellia, M. chitwoodi, M. deconincki, M. exigua, M. graminis, M. hapla, M. hispanica, M. incognita, M. javanica, M. kirjanovae, M. kralli, M. litoralis, M. lusitanica, M. maritima* and *M. naasi* were compared, available living material was used for morphological study and isozyme tests.

It is concluded that the original species descriptions of *M. deconincki* and *M. litoralis* were based on a mixture of *M. ardenensis* and *M. hapla*. Both are considered as junior synonyms of *M. ardenensis*. No types or populations of *M. megriensis* were available, it is regarded as a *species inquirenda*. The occurrence of *M. exigua* and *M. graminis* was reviewed, it is not likely that either is native to Europe. *Meloidogyne kirjanovae* is considered as a junior synonym of *M. incognita*. The status of the species described by Chitwood (1949) is discussed in historical perspective.

An extensive description of *M. fallax*, a relatively new species, close related to *M. chitwoodi* was added. Types of *M. maritima* were not available, samples from the UK type locality included *M. maritima* and an undescribed root-knot nematode. *M. maritima* was redescribed and new types prepared and deposited. The unknown species was described as *M. duytsi*. Both species parasitize on different coastal dune grasses. Coastal foredune samples from several location in England, Wales, the Netherlands, Belgium, France and Portugal (one location), all included both species. They may be considered as the most widely distributed root-knot nematodes in European natural coastal areas. Another undescribed species, resembling *M. caraganae* and *M. turkestanica*, was found near the river Rhine in Germany and the Netherlands, together with *M. ardenensis* and *M. hapla*. The description is retarded because of the noticed species complex. The river Rhine could play an important role in the distribution of root-knot nematodes.

Several new esterase and malate dehydrogenase isozyme types were detected within the European root-knot nematodes. The intraspecific isozyme variability is very low and therefore useful for species identification. Therefore identification keys based on isozyme types and morphological characters were prepared.

Perineal pattern

The perineal pattern, a unique and complex character located at the female posterior body region, comprises the vulva-anus area, tail terminus, phasmids, lateral lines and surrounding cuticular striae. Little is known about the development of this important but variable character. Literature data added with morphological observations on female development and perineal patterns were used to explain in detail the perineal pattern.

The basic pattern, with fine striae, is formed just after the last moult when the vulva is induced above the anus at the ventral body side. This proces is likely to be influenced by the previous feeding stage. The vulva-anus region moves posteriorly, while the female is still enclosed by the old cuticle layers. After the female starts feeding again, she increases rapidly in width, pattern striae become folded and coarser. The role of the expanding rectal glands and perineal pattern structures as lateral lines, anus, tail terminus and punctations are discussed in detail. Vulva-anus areas in other related nematode genera were observed, it is concluded that perineal patterns have been depicted upside-down during the last fifty years.

Appendix I

The nominal root-knot nematode species (*Meloidogyne* Göldi, 1892).

Type species

M. exigua Göldi, 1892
  syn. *Heterodera exigua* (Göldi, 1892) Marcinowski, 1909

Species

M. acronea Coetzee, 1956
  syn. *Hypsoperine acronea* (Coetzee, 1956) Sledge & Golden, 1964
    Hypsoperine (Hypsoperine) acronea (Coetzee, 1956) Siddiqi, 1985
M. actinidiae Li & Yu, 1991
M. africana Whitehead, 1960
M. aquatilis Ebsary & Eveleigh, 1983
M. arabicida López & Salazar, 1989
M. ardenensis Santos, 1968
  syn. *M. deconincki* Elmiligy, 1968
    M. litoralis Elmiligy, 1968
M. arenaria (Neal, 1889) Chitwood, 1949
  syn. *Anguillula arenaria* Neal, 1889
    Heterodera arenaria (Neal, 1889) Marcinowski, 1909
    M. arenaria arenaria (Neal, 1889) Chitwood, 1949
    M. arenaria thamesi Chitwood, 1952
    M. thamesi (Chitwood, 1952) Goodey, 1963
M. artiellia Franklin, 1961
M. brevicauda Loos, 1953
M. californiensis Abdel-Rahman & Maggenti, 1987
M. camelliae Golden, 1979
M. caraganae Shagalina, Ivanova & Krall, 1985
M. carolinensis Eisenback, 1982
M. chitwoodi Golden, O'bannon, Santo & Finley, 1980
M. christiei Golden & Kaplan, 1986
M. cirricauda Zhang, 1991
M. citri Zhang, Gao & Weng, 1990
M. coffeicola Lordello & Zamith, 1960
  syn. *Meloidodera coffeicola* (Lordello & Zamith, 1960) Kirjanova, 1963
M. cruciani Garcia-Martinez, Taylor & Smart, 1982
M. cynariensis Bihn, 1990
M. decalineata Whitehead, 1968
M. donghaiensis Zheng, Lin & Zheng, 1990
M. duytsi Karssen, van Aelst & van der Putten, 1998
M. enterolobii Yang & Eisenback, 1983
M. ethiopica Whitehead, 1968

M. fallax Karssen, 1996

M. fanzhiensis Chen, Peng & Zheng, 199O

M. fujianensis Pan, 1985

M. graminicola Golden & Birchfield, 1965

M. graminis (sledge & Golden, 1964) Whitehead, 1968

  syn. *Hypsoperine graminis* Sledge & Golden, 1964

      Hypsoperine (Hypsoperine) graminis (Sledge & Golden, 1964) siddiqi, 1985

M. hainanensis Liao & Feng, 1995

M. hapla Chitwood, 1949

M. hispanica Hirschmann, 1986

M. ichinohei Araki, 1992

M. incognita (Kofoid & White, 1919) Chitwood, 1949

  syn. *Oxyuris incognita* Kofoid & White, 1919

      Heterodera incognita (Kofoid & White, 1919) Sandground, 1923

      M. incognita incognita (Kofoid & White, 1919) Chitwood, 1949

      M. incognita acrita Chitwood, 1949

      M. acrita (Chitwood, 1949) Esser, Perry & Taylor, 1976

      M. incognita inornata Lordello, 1956

      M. inornata Lordello, 1956

      M. kirjanovae Terenteva, 1965

      M. elegans da Ponte, 1977

      M. grahami Golden & Slana, 1978

      M. incognita wartellei Golden & Birchfield, 1978

M. indica Whitehead, 1968

M. javanica (Treub, 1885) Chitwood, 1949

  syn. *Heterodera javanica* Treub, 1885

      Anquillula javanica (Treub, 1885) Lavergne, 19O1

      M. javanica javanica (Treub, 1885) Chitwood, 1949

      M. javanica bauruensis Lordello, 1956

      M. bauruensis (Lordello, 1956) Esser, Perry & Taylor, 1976

      M. lucknowica Singh, 1969

      M. lordelloi da Ponte, 1969

M. jianyangensis Yang, Hu, Chen & Zhu, 199O

M. jinanensis Zhang & Su, 1986

M. kikuyensis de Grisse, 196O

M. konaensis Eisenback, Bernard & Schmitt, 1994

M. kongi Yang, Wang & Feng, 1988

M. kralli Jepson, 1984

M. lini Yang, Hu & Xu, 1988

M. lusitanica Abrantes & Santos, 1991

M. mali Itoh, Oshima & Ichinohe, 1969

M. maritima (Jepson, 1987) Karssen, van Aelst & Cook, 1998

M. marylandi Jepson & Golden, 1987

M. mayaguensis Rammah & Hirschmann, 1988

M. megadora Whitehead, 1968

M. megatyla Baldwin & Sasser, 1979

M. mersa Siddiqi, 1992

M microcephala Cliff & Hirschmann, 1984

M. microtyla Mulvey, Townshend & Potter, 1975
M. mingnanica Zhang, 1993
M. morocciensis Rammah & Hirschmann, 1990
M. naasi Franklin, 1965
M. nataliei Golden, Rose & Bird, 1981
M. oryzae Maas, Sanders & Dede, 1978
M. oteifae Elmiligy, 1968
M. ottersoni (Thorne, 1969) Franklin, 1971
   syn. *Hypsoperine ottersoni* Thorne, 1969
      Hypsoperine (Hypsoperine) ottersoni (Thorne, 1969) Siddiqi, 1985
M. ovalis Riffle, 1963
M. paranaensis Carneiro, Carneiro, Abrantes, Santos & Almeida, 1996
M. partityla Kleynhans, 1986
M. pini Eisenback, Yang & Hartman, 1985
M. plantani Hirschmann, 1982
M. propora Spaull, 1977
   syn. *Hypsoperine (Hypsoperine) propora* (Spaull, 1977) Siddiqi, 1985
M. querciana Golden, 1979
M. salasi Lopez, 1984
M. sasseri Handoo, Huettel & Golden, 1993
M. sewelli Mulvey & Anderson, 1980
M. sinensis Zhang, 1983
M. spartinae (Rau & Fassuliotis, 1965) Whitehead, 1968
   syn. *Hypsoperine spartinae* Rau & Fassuliotis, 1965
      Hypsoperine (Hypsoperine) spartinae (Rau & Fassuliotis, 1965) Siddiqi, 1985
M. subartica Bernard, 1981
M. suginamiensis Toida & Yaegashi, 1984
M. tadshikistanica Kirjanova & Ivanova, 1965
M. turkestanica Shagalina, Ivanova & Krall, 1985
M. trifoliophila Bernard & Eisenback, 1997
M. triticoryzae Gaur, Saha & Khan, 1993
M. vandervegtei Kleynhans, 1988

Species inquirendae

M. megriensis (Poghossian, 1971) Esser, Perry & Taylor, 1976
   syn. *Hypsoperine megriensis* Poghossian, 1971
      Hypsoperine (Hypsoperine) megriensis (Poghossian, 1971) Siddiqi, 1985
M. marioni (Cornu, 1879) Chitwood & Oteifa, 1952
   syn. *Anguillula marioni* Cornu, 1879
      Heterodera marioni (Cornu, 1879) Marcinowski, 1909
      Heterodera goeldi Lordello, 1951 = nom. nov. for *M. marioni*
M. vialae (Lavergne, 1901) Chitwood & Oteifa, 1952
   syn. *Anquillula vialae* Lavergne, 1901
      Heterodera vialae (Lavergne, 1901) Marcinowski, 1909
M. poghossianae Kirjanova, 1963
   syn. *M. acronea* apud Poghossian, 1961

Appendix II

Identification keys to European root-knot nematodes

Morphology

1. Perineal pattern with punctations in tail terminus area, J2 with irregular
    shaped hyaline tail terminus (clubbed to bifid)...*M. hapla*
    Perineal pattern without punctations and irregular J2 tail terminus............................2.

2. J2 hemizonid posterior to S-E pore...3.
    J2 hemizonid anterior or at the same level as the S-E pore..4.

3. J2 tail ≤ 45 μm...*M. ardenensis*
    J2 tail > 45 μm...*M. maritima*

4. Perineal pattern with distinct lateral lines clearly separating dorsal and
    ventral parts...*M. javanica*
    Perineal pattern without distinct lateral lines...5.

5. J2 and males with several small vesicles in the anterior metacorpus.................*M. naasi*
    J2 and males without vesicles in the anterior metacorpus..6.

6. Perineal pattern clearly trapezoidal with coarse striae................................*M. lusitanica*
    Perineal pattern not trapezoidal..7.

7. Male stylet ≥ 21 μm...8.
    Male stylet < 21 μm...10.

8. Male stylet knobs backwardly sloping, DGO ≥ 4 μm................................*M. arenaria*
    Male stylet knobs set off, DGO < 4 μm..9.

9. Male labial disk raised above medial lips..*M. incognita*
    Male head cap rounded..*M. hispanica*

10. J2 tail ≤ 55 μm...11.
    J2 tail > 55 μm...13.

11. Male stylet knobs rounded and set off...*M. fallax*
    Male knobs oval to irregular and backwardly sloping...12.

12. Perineal pattern rounded to oval...*M. chitwoodi*
    Perineal pattern small, dorsal arch high, angular with coarse and thick
    striae near perineum...*M. artiellia*

13. J2 tail tip sharply pointed..*M. kralli*
    J2 tail tip rounded..*M. duytsi*

Isozymes

The relative isozyme movement (Rm %) of non-specific esterase (Est) and malate dehydrogenase (Mdh)*.

Species	Mdh		Est	
M. artiellia	N1b	**20**	M2-VF1	**61, 63½, 73½**
M. fallax	N1b	**20**	F3	31, 40, 58
M. arenaria	N1	**28**	A3	**58, 61, 63½**
			A2	**61, 63½**
M. hispanica	N1	**28**	S2-M1	**47, 51, 54½**
M. incognita	N1	**28**	I1	**54½**
M. javanica	N1	**28**	J3	**54½, 62, 65½**
M. ardenensis	N1a	**31½** (+ 43)	-	28, 44
M. chitwoodi	N1a	**31½**	S1	**51**
M. naasi	N1a	**31½**	VF1	**73½**
M. duytsi	N1c	**41** (+ 37½)	VS1	**38**
M. kralli	N1c	**41**	-	48, 85, 89
M. lusitanica	N1c	**41** (?)	A1	**63½**
M. maritima	N1c	**41**	VS1-S1	**38, 49**
M. hapla	H1	**47½**	H1	**57½**

* isozyme coding according to Esbenshade & Triantaphyllou (1985), weakly stained types or bands are not printed in bold type. The N1c esterase type of *M. lusitanica* is uncertain and needs confirmation.

NEMATOLOGY
International Journal of Fundamental and Applied Nematological Research

EDITOR-IN-CHIEF:
Roger Cook, *Institute of Grassland and Environmental Research, Aberystwyth, Ceredigion SY23 3EB, Wales, UK.*
David J. Hunt, *Cabi Bioscience, UK Centre, Egham, Surrey, TW20 9TY, UK.*

AIMS AND SCOPE
Nematology is an international journal for the publication of all aspects of nematological research, from molecular biology to field studies. Papers on nematode parasites and anthropods, on soil free-living nematodes, and on interactions of these and other organisms, are particularly welcome. Research on fresh water and marine nematodes is also considered when the observations are of more general interest.

Nematology publishes full research papers, short communications, foresight articles (which permit an author to express a view on current or fundamental subjects), abstracts of nematology meetings, full communications at nematology symposia, reviews of books and other media, announcements of nematological activities and societies.

ISSN (print) 1388-5545 • ISSN (online) 1568-6411
2002: Volume 4 in 8 issues • Subscription price: EUR 480 / US$ 570 –
including access to online edition

Subscription order form/Sample copy request
Send orders to your usual agent or to:
Brill Academic Publishers, P.O. Box 9000, 2300 PA Leiden, The Netherlands
Tel: +31 71 535 3500 Fax: +31 71 531 7532 E-mail: orders@brill.nl
North America
Brill Academic Publishers, P.O. Box 605, Herndon, VA 20172
Tel: 1-800-337-9255. Fax: 1-703-661-1501 E-mail: cs@brillusa.com

Printed in the United States
By Bookmasters